Sustainable Engineering for Life Tomorrow

Environment and Society

Series Editor: Douglas Vakoch

As scholars examine the environmental challenges facing humanity, they increasingly recognize that solutions require a focus on the human causes and consequences of these threats, and not merely a focus on the scientific and technical issues. To meet this need, the Environment and Society series explores a broad range of topics in environmental studies from the perspectives of the social sciences and humanities. Books in this series help the reader understand contemporary environmental concerns, while offering concrete steps to address these problems.

Books in this series include both monographs and edited volumes that are grounded in the realities of ecological issues identified by the natural sciences. Our authors and contributors come from disciplines including but not limited to anthropology, architecture, area studies, communication studies, economics, ethics, gender studies, geography, history, law, pedagogy, philosophy, political science, psychology, religious studies, sociology, and theology. To foster a constructive dialogue between these researchers and environmental scientists, the Environment and Society series publishes work that is relevant to those engaged in environmental studies, while also being of interest to scholars from the author's primary discipline.

Recent Titles in the series

Sustainable Engineering for Life Tomorrow

Edited by
Jacqueline A. Stagner and David S. K. Ting

LEXINGTON BOOKS
Lanham • Boulder • New York • London

Published by Lexington Books
An imprint of The Rowman & Littlefield Publishing Group, Inc.
4501 Forbes Boulevard, Suite 200, Lanham, Maryland 20706
www.rowman.com

6 Tinworth Street, London SE11 5AL, United Kingdom

British Library Cataloguing in Publication Information Available

Library of Congress Cataloging-in-Publication Data

ISBN 978-1-7936-2501-4 (cloth)
ISBN 978-1-7936-2503-8 (pbk)
ISBN 978-1-7936-2502-1 (electronic)

To everyone who engineers and lives sustainably.

Contents

Preface

According to Thomas Sigsgaard, "The most sustainable way is to not make things. The second most sustainable way is to make something very useful, to solve a problem that hasn't been solved." This should motivate intelligent beings to actively embark on progressing through problem-solving for a more sustainable tomorrow. Incidentally, Henry Ward Beecher correctly defined progress when he proclaimed, "We should live and labor in our time that what came to us as a seed may go to the next generation as blossom, and what came to us as blossom, may go to them as fruit. This is what we mean by progress." *Sustainable Engineering for Life Tomorrow* is a collection of striving toward this progress.

What do we do with the aging infrastructures that seem to be out of place and time? Al-Kodmany brings to life such an abandoned infrastructure in one of the biggest megacities, Chicago, in chapter 1, "Retrofitting Aging Infrastructure: The Case of Chicago Riverwalk." Specifically, he details the transformation of an outmoded infrastructure and disused river's banks into an attractive gathering civic space, a linear urban park, and a functional transportation corridor. Talking about old infrastructures, Kou and Zhong expound on much-needed advanced window retrofitting technologies in chapter 2, "Sustaining Old Infrastructures with Advanced Window Retrofitting Technologies." It is difficult for many to accept that windows account for four quads (4×10^{18} J) of primary energy loss every year in the United States alone. The potential to save energy and money worldwide by taking care of the window is huge! System thinking that considers the big picture is needed to further sustain smart cities. Minaei presents this comprehensively in chapter 3, "A Critical Review of Urban Energy Solutions and Practices." The key is to execute focused improvements while keeping the entire system in mind. To put available wind into usage, Balo, Yilmaz, and Sua disclose "Energy

Efficiency and Sustainability through Wind Power for Green Hospitals" in chapter 4. Energy-intensive hospitals necessitate continuous, round-the-clock energy supply. This automatically calls for reliable energy storage. Depending on specific requirements, few technologies can compete with a "Supercapacitor for Sustainable Energy Storage," as brought to light by Zhang, Shang, and Yi in chapter 5. Two unique strengths make supercapacitors stand out; these are their burst charging and releasing speed and their extreme power density. Progress in energy storage technologies will ultimately further the harnessing of solar energy. Coelho, Schmitz, and Martins present the basic topics related to photovoltaic energy generation in chapter 6, "Introduction to Photovoltaic Energy Generation." This comprehensive chapter contains everything that a solar enthusiast needs for basking in solar energy. What about learning from chemotaxis, directed cell motion subjected to a changing environment? Readers are encouraged to random walk into chapter 7, "Functional-form Sufficiency Achieved by Biased Random Walk: Behavioral Model in Architectural Bioactive Design." Let Abdallah, Estévez, Khalil, Tantawy, and Sobhy enlighten you with their aesthetic, bioactive pavilions for producing bioelectricity in microbial fuel cells. Baker, Tinggaard, Enevoldsen, and Xydis extend a circular economy strategy, such as that applied in the automotive industry, to the kitchen manufacturing business in chapter 8, "Circularity in Kitchen Design and Production Business: A Sustainable and Disruptive Model." The application of leasing and trade-in can attract and retain customers with a significant fringe benefit of bettering large-item waste management and reduction. The value of a dollar intrinsically encompasses regional politics, policies, and cultures. This is illustrated by a timely example on offshore wind development by Nichol, Miller, and Carriveau, in chapter 9, "Should Canada Pursue or Support Offshore Wind Development?" Their findings bespeak exporting Canadian offshore oil and gas expertise to support international offshore wind development as the most favorable scenario. It is thus clear that judicious decisions must be based on appropriate considerations of all significant factors involved. Gökgöz and Yalçın invoke multi-criteria decision-making (MCDM) to evaluate renewable energy sources in chapter 10, "Analyzing the Renewable Energy Sources of Nordic and Baltic Countries with MCDM Approach." Their MCDM analysis reveals the underlying reasons behind the strategic breakthroughs realized by Nordic and Baltic countries in simultaneous climate change mitigation and economic gain, setting an enviable example for the rest of the world to follow.

Acknowledgments

Michael Gibson, the then senior acquisitions editor of Lexington Books, a division of Rowman & Littlefield Publishing Group, gallantly sealed the contract to proceed with *Sustainable Engineering for Life Tomorrow* on November 14, 2019. The circumstances brought about by Covid-19 seriously hindered this endeavor, which was supposed to be wrapped up by October 1, 2020. Thankfully, providence came in the nick of time. Without knowing anything about the book editors who she had to work with, Kasey Beduhn "adopted" us with no questions asked. Needless to say, the entire amazing Lexington Books team followed her lead "blindly." With the wind behind our back, the last lap went smoothly. Not least of all, we are grateful to all the contributors, experts who compiled the chapters, and reviewers who furthered their quality. Thank you for journeying with us.

Jacqueline A. Stagner and David S. K. Ting
Turbulence & Energy Laboratory
University of Windsor

Chapter 1

Retrofitting Aging Urban Infrastructure

The Case of Chicago Riverwalk

Kheir Al-Kodmany

INTRODUCTION

Sustainability

Undoubtedly, the concept of urban sustainability has been useful in supporting urban developments. In 2015, the United Nations adopted the 2030 Agenda for Sustainable Development which details 17 Sustainable Development Goals (SDGs) and 169 Actionable Targets to be realized by 2030. In particular, Goal #11 refers to creating sustainable cities and communities. Further, the United Nations' World Urban Forum (WUF), the world's premier conference on urban development, has embraced "sustainability" as an overarching theme for its agendas. The commitment to SDGs has been apparent since WUF's first meeting in 2002, titled "Sustainable Urbanization," in Nairobi, Kenya, through the latest in 2020, in Abu Dhabi, United Arab Emirates. In the same vein, in 2016, the United Nations Conference on Housing and Sustainable Urban Development (Habitat III) adopted the New Urban Agenda (translated to thirty-three languages), which also stresses sustainability. Similar to the United Nations' focus on and interest in sustainability, other important organizations, such as the World Bank, the Global Environment Facility (GEF), Local Government for Sustainability (ICLEI), and Global Platform for Sustainable Cities (GPSC) have worked on and supported local and global sustainability projects, initiatives, and programs (United Nations).

Likewise, the term "sustainability" frequently appears in academic literature and is discussed in professional conferences. In the United States, the American Planning Association (APA), the prime professional planning

1

organization, continues to use the term "sustainability" in its National Planning Conference (NPC) and publications. In 2010, at the United Nations' fifth WUF, the APA announced the creation of the *Sustaining Places Initiative*, a program that focuses on sustainability as a key to all urban planning activities. In recent years, the program has published several key reports, articles, and books that highlight this planning approach; see for example, *Sustaining Places: Best Practices for Comprehensive Plans* by David R. Godschalk and David C. Rouse (2015).

As early as 1987, the United Nations crafted a profoundly useful definition of sustainability, stating it as "meeting the needs of the present without compromising the ability of future generations to meet their own needs" (WCED, 1987). Owing to its concise and precise nature, this definition continues to be among the most frequently used definitions. The United Nations' definition coincides with the core meaning of the term "sustain," which refers to long-term nourishment, strengthening, and support (e.g., Merriam-Webster Dictionary). In addition, the United Nations' definition aligns with the planning profession's focus on future conditions while considering that resources are finite. Sustainability's three Rs (i.e., Reduce, Reuse, and Recycle) comfortably fit the urban planning profession's goals and objectives (Al-Kodmany, 2018).

Notably, sustainability provides a holistic framework to illustrate the connection between three important urban dimensions, namely the social, economic, and environmental, also known as the "3 Ps" that is, people, profit, and the planet, where:

- "people" (social) refers to community welfare, equity, inclusiveness, livability, and residents' health;
- "profit" (economic) refers to growth, employment, wages, revenue, prosperity, trade, and global competitiveness; and
- "planet" (environment) refers to improving environmental health, conserving and preserving ecosystems, using natural resources efficiently, enhancing infrastructure, managing traffic, reducing greenhouse gas (GHG) emissions, mitigating the impacts of climate change, and adapting to climate change.

These three sustainability dimensions are also articulated by "3 Es" of equality (social), economics (financial), and ecology (environment). Both in academic literature and practical world, the three Ps or Es are referred to as the triple bottom line (TBL). Importantly, "Sustainability seeks to *balance* these three dimensions across geographic scales—from individual habitats to neighborhood, community, city, region, country, continent, and the planet at large—and according to both short- and long-term goals" (Al-Kodmany, 2018, p. 3).

In the context of this research, it is argued that sustainability provides a useful framework to design, retrofit, and evaluate infrastructure improvement projects. The sustainability thesis helps to examine and improve the physical space, social life, and economic conditions, which collectively enhances the overall quality of life of the city. A dynamic relationship harnesses the synergy of sustainability's three dimensions. That is, attractive public spaces invite more local people and tourists. In turn, tourists increase demand on local businesses, social amenities, and services (e.g., restaurants, cafes, hotels, travel agencies, logistics, media, banks, and the like); thereby, igniting social life, creating jobs, enhancing the economy, and improving the environment.

Sustainability Projects in Chicago

Chicago has been a world leader in urban sustainability. Commitment to sustainability has been stressed by at least three consecutive mayors of the city, Richard M. Daley (1989–2011), Rahm Emanuel, (2011–2019), and Lori Lightfoot (2019–). For example, Rahm Emanuel stated "Promoting sustainability throughout the city is a key focus of my administration, and is integral to the quality of life we are working to develop for all Chicagoans. I want Chicago to be the greenest city in the world, and I am committed to fostering opportunities for Chicagoans to make sustainability a part of their lives and their experience in the city" (Office of the Mayor, 2012).

Chicago has been pioneering several sustainable and green retrofit infrastructure projects. For example, through the Green Alleys initiative, the city has been retrofitting alleys' pavement into permeable surfaces to decrease basement flooding and sewer run-off. A green alley absorbs 80% of stormwater that falls and passes over it. Since 2006, the city has completed over 200 green alleys. The city has also been working on replacing "900 miles [1,448 kilometers] of century-old water mains and 275 miles [443 kilometers] of sewer mains and the lining of 160,000 catch-basins and 700 miles [1,127 kilometers] of sewer mains" (Sustainable Chicago, 2015, p. 22). Chicago is also rebuilding tens of miles of aging Chicago Transit Authority (CTA) train tracks. Recently, it has embarked on a project to replace underground gas pipes (Henchen and Kroh, 2020). Among the ambitious infrastructure projects that the city has accomplished is retrofitting a portion of the Chicago River.

The Chicago River

The Chicago River is one of the greatest assets in the Chicago region and State of Illinois. To begin with, it was an essential element to Chicago's existence and growth. It has supported agricultural, trade, commerce, and industrial and transport activities; and, as such, it has attracted many residents, workers, and visitors. As early as the 1600s, the river and Lake Michigan served as major trade routes. When Chicago was incorporated in 1836, large orchards and agricultural fields began to grow around the river, which provided the city's inhabitants with their basic needs. Also, sizable manufacturing places and industrial complexes (e.g., slaughterhouses, stockyards, meatpacking plants, tanneries, and lumber and steel mills) quickly proliferated in the area. Unfortunately, the river served as a sewer to the increasing population and industrial activities, causing serious pollution of the river for over a century. Consequently, the river became derelict, neglected, and eyesore that discouraged new developments and disinvited people.

In 1900, Chicago built the Chicago Sanitary and Ship Canal (CSSC) which provided transportation and sanitary functions. That is, by building the canal deeper and deeper as it progressed west, the canal reversed the flow of the Main Stem and South Branch of the Chicago River, drawing its water away from Lake Michigan. This extraordinary project solved severe water pollution problems in the lake and river that were cause by dramatic population increase and rapid industrial developments that dumped their waste in the river. The CSSC kept sewage out of the city's drinking supply (Lake Michigan) and flushed the filthy Chicago River with clean Lake Michigan water (Bosch, 2008; Dyja, 2014; Smith, 2006).

Notably, city planners Daniel Burnham and Edward H. Bennett made the Chicago River a focal point of their 1909 Plan of Chicago. They envisioned making Chicago the "Paris of the Prairie" and to that end they suggested several critical infrastructure systems and "beautification" projects, including widening roads, integrating new parks, and building civic places and cultural amenities. The 1909 Plan also aimed to alleviate the overcrowding of ships in the narrow river by building several lakefront piers. Out of that proposal came Navy Pier, completed in 1916. Burnham and Bennett also envisioned a magnificent bridge to join urban areas south and north of the Chicago River at Michigan Avenue, already Chicago's main street. The double-decker Michigan Avenue Bridge (DuSable Bridge) was completed in 1920, and the double-decker Wacker Drive was completed in 1926. The plan also called for an elegant esplanade lining the river's Main Stem (Smith, 2006).

By the late 1960s, sewer overflows became excessive and the United States Environmental Protection Agency (USEPA) demanded that the Metropolitan Water Reclamation District (MWRD) work harder to clean up the river. In the early 1970s, the passage of the federal Clean Water Act led to planning for the Deep Tunnel system (also known as Tunnel and Reservoir Plan (TARP)), which is designed to hold waste and runoff until it can be safely treated. Construction of the project started in 1975, and it is anticipated to be completed in 2029. TARP should be capable of holding up to 20.55 billion gallons of excess water (Hunt and DeVries, 2017).

In parallel to governmental efforts, several local environmental groups (e.g., Friends of the Chicago River, FOCR) voiced concerns about the deteriorating river condition. In collaboration with other groups and agencies, FOCR has been organizing volunteers to remove invasive plants and seed the riverbanks with native, local species. Consequently, these organizations have not only reduced the amount of runoff that flows into the river but also have restored dozens of acres of fish habitats along the river. In 1998, the City of Chicago worked with FOCR to release a plan for the development of the Chicago Riverwalk, a continuous river trail that allows the city's denizens and visitors to enjoy the river (Johnson, 2006).

Fortunately, over the past two decades, the city, local residents, community-based organizations, and politicians have exerted serious efforts to improve the river's conditions. As the river passes through the downtown and touches thirty-three of the city's seventy-seven communities, the river's conditions immediately affect the quality of life and the sustainability of the city. Improving the river's conditions is much needed to enhance ecological conditions, restore and support aquatic species, invite residential developments, encourage economic development, spur recreational opportunities, and attract tourists. Consequently, the City of Chicago had embarked on building the Riverwalk. It is touted as the city's "Second Waterfront," after the Lakefront, which was saved by Chicagoan visionaries and leaders since the mid-1800s.

The Chicago Riverwalk

The Riverwalk is a result of a long-term vision, sustained efforts, and earnest commitments among numerous stakeholders and agencies. Between 1990 and 1995, the Chicago Department of Transportation (CDOT) conducted a feasibility study and produced planning guidelines for the development of the main branch of the river. In the late 1990s, the City of Chicago worked with Skidmore, Owings, and Merrill (SOM) to set a plan for an area that extends from Lakeshore Drive to Michigan Avenue. Later, in the early 2000s, CDOT submitted a request to the U.S. Department of Transportation to permit building out into the river. Also, during this time, the Chicago Park District completed a schematic master plan and CDOT started collaborating with Ross Barney Architects on Phase 1, which they completed in 2009. Next, in 2011, Sasaki and Ross Barney Architects started working on Phases 2 and 3, which they completed in 2015 and 2016, respectively. In the meantime, in 2013, the City of Chicago received a Transportation Infrastructure Finance and Innovation Act (TIFIA) loan of $98,660,000 for constructing Phases 2 and 3. Notably, the project received support from Mayors Richard M.

Daley (1989–2011) and Rahm Emanuel (2011–2019) and various nonprofit organizations, particularly FOCR and the Chicago Loop Alliance, as well as other agencies such as the Chicago Park District, the Department of Fleet and Facility Management (2FM), and the Department of Cultural Affairs and Special Events (DCASE).

In addition to improving the river's condition, an important trigger of the Riverwalk project was the reconstruction of Wacker Drive, a major multitier thoroughfare that serves the downtown and runs alongside two main branches of the Chicago River. Constructed in 1920s, the conditions of this seventy-year-old major thoroughfare began deteriorating and crumbling, creating traffic jams in the downtown. The reconstruction of the Wacker Drive was carried out into two phases. The rebuilding of the east-west section of the thoroughfare was completed in 2002 and the south-west segment was completed in 2012. Importantly, the design and construction of the new double-decker road took in consideration the retrofitting process of the Riverwalk project. The reconfigured sidewalks and plazas on the east-west of the Wacker Drive were designed to support greater public access to the Riverwalk; for example, ensuring seamless access between Wacker Drive sidewalk and the River Theater, see section 2.3.3. In a nutshell, looking at the process positively, the reconstruction of Wacker Drive stimulated improvement along the river and the construction of the Riverwalk.

The planning vision focused on using the Riverwalk project as a means to reclaim the Chicago River to improve ecological conditions, promote economic growth, enhance health, and foster happiness, well-being, and safety. The project initially was built on the architect and urban planner Daniel Burnham's vision of creating a pedestrian civic promenade for the river, presented in the early 1900s, following the celebrated reversal of the river. As such, the revived vision intended to attract residents and visitors down to the river level so that they enjoy and appreciate the natural environment of the city. Another important goal was to create comfortable and accessible pedestrian links in Chicago Downtown. By connecting Chicago's "Second Waterfront" (Riverwalk Waterfront) with the "First Waterfront" (Lake Michigan Waterfront), much of the critical elements of the downtown become accessible on foot. The project's goals can be summarized as follows:

- Transform underutilized waterfront infrastructure into a sustainable amenity.
- Enhance downtown by giving residents and tourists access to a wide spectrum of riverfront recreational activities.
- Build an uninterrupted, car-free promenade along the south bank of the Main Branch, from the Lakefront to Lake Street.
- Establish intermodal connections among bike-share stations and public mass transit nodes (bus, train, and boat).

- Improve the river's water quality and restore its ecology as a valuable natural resource.
- Support local businesses and create new ones.

The Federal Government designates the Chicago River as "Waters of the United States." Ergo, any construction or changes to its path would require approval from the U.S. Congress. Several agencies and organizations (e.g., Friends of the River, the United States Coast Guard, Illinois Department of Natural Resources, the Volpe Transportation Institute, among others) crafted a plan that proposed building out into the river to accommodate the Riverwalk, thereby altering the 200-foot-wide (61-meter-wide) navigational channel. In 2004, the U.S. Congress permitted the city to build out into the river, as follows:

- 20 feet (6 meters) at the areas right beneath each of the six Bascule Bridges.
- 25 feet (7.6 meters) at the areas between the bridges.
- 50 feet (15.2 meters) at the far western end of the path, near the river confluence.

DESCRIPTION OF THE CHICAGO RIVERWALK

The Chicago Riverwalk is a public walkway along the south bank of the Main Branch of the Chicago River. By retrofitting its disused docks and building under-bridge crossings, today, it is a continuous promenade that connects the Lakefront (near the river mouth) with Lake Street (near the river confluence, right at the western edge of the Chicago Loop). This 1.25-mile (2-kilometer) stretch is a crucial spine that knits together many downtown places, including the Loop, Navy Pier, Magnificent Mile (Mag Mile), and other important amenities and tourist attractions such as Millennium Park and Maggie Daley Park. Various studies that engaged the Riverwalk, divide its area and construction process into four parts:

1) Recreational Trail,
2) Phase 1,
3) Phase 2, and
4) Phase 3

Recreational Trail

The Recreational Trail or Riverwalk Esplanade extends from Lake Shore Drive to Michigan Avenue. Completed in early 2000s, this "east part" is a pedestrian and bike path and a docking place for boats that offer educational and recreational tours. It features lush landscaping, trees, plants, play areas, attractive LED lighting fixtures, furniture, railing, seating areas, and public restrooms. Efforts for retrofitting and improving this area started in early 2000s and continue to the present (figure 1.1).

The "west part" of the Riverwalk extends from Michigan Avenue to Lake Street. While the "east part" emphasizes on a natural park-like environment, the "west part" evokes the image of a civic space. Interestingly, the west part consists of block-long sections or "bays" that contain civic plazas. It was constructed in three phases.

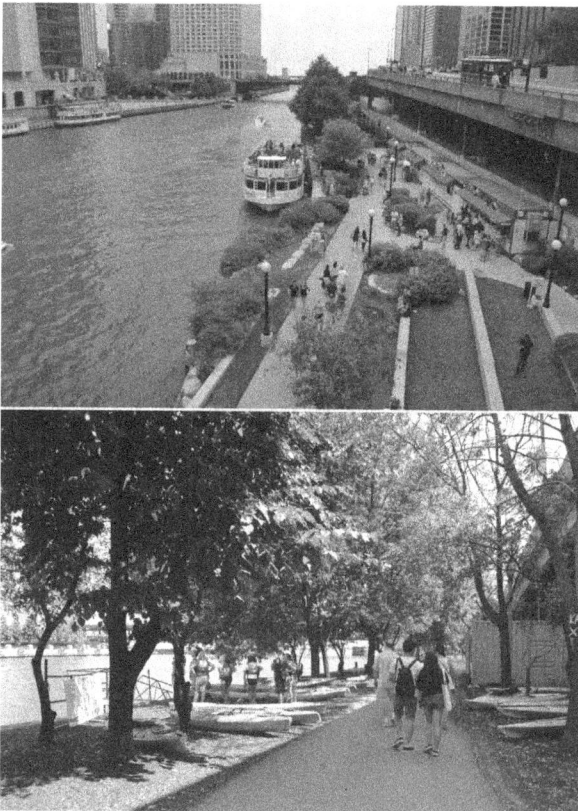

Figure 1.1   The Recreational Trail, the "east part" of the Riverwalk. It intends to provide a transition from a natural park (near Lake Michigan) to a civic space (near Michigan Avenue). The top photo offers an overview of the Recreational Trail; note the tourists' riverboat docking area. The bottom photo shows the trail part closer to the Lakefront. *Source*: Photo by the author.

Phase 1

Phase 1 extends from Michigan Avenue to State Street. Completed in 2009, it contains the McCormick Bridgehouse & Chicago River Museum, the Chicago's Vietnam Veterans Memorial, and two pedestrian connections at the river level. These pedestrian crossings were important to ensure continuity of the path. The older situation was restricting and discouraging. Pedestrians had to ascend stairways to the upper-level of Wacker Drive, then cross a city street, and later descend again to the river level. Therefore, under-bridge crossings enabled pedestrians to traverse along the river uninterruptedly.

McCormick Bridgehouse & Chicago River Museum

Designed by architect and urban planner Edward H. Bennett, the McCormick Bridgehouse & Chicago River Museum is a five-story, 1,200 sq. feet (111 sq. meter) bridgehouse located at the western edge of Michigan Avenue. At the river level, this historic structure lively exhibits the bridge's machinery and massive gears that make it open and close. It also exhibits a series of visuals that illustrate the story of eighteen moveable bridges that dot the river. Further, exhibits explain historic events; for example, the Eastland Disaster event that took place in the nearby area between Clark and LaSalle streets. The museum's top level allows visitors to enjoy 360-degree views of the city. "Cozy" interior spaces feature exposed brick walls and steel beams. In contrast, the exterior features a 1928 bas-relief by Henry Hering, titled "The Defenders." It commemorates the 1812 Battle of Fort Dearborn (Chicago Architecture Center) (figure 1.2).

Vietnam Veterans Memorial

Completed in 2005, Chicago's Vietnam Veterans Memorial occupies the block between Wabash and State streets. It encompasses multiple tiers that contain several water features, including a waterfall, a reflecting pool, and a fountain with fourteen jets. Along the waterfall, the names of all late Illinois Vietnam War veterans are engraved in a long granite slab. It is one of the largest Vietnam War Memorials outside of Washington, D.C. The memorial is heavily planted and is secluded from the street, providing a refuge from the noise and activities of the city. The Vietnam Veterans Memorial's design produces a quiet and peaceful space, and yet many walkers and joggers pass through it, with some joggers running up and down the stairs of the plaza as a part of their routine. Many photographers can also be seen taking pictures of the spectacular river environments and surrounding skyscrapers. The most popular spaces are in or near the grassy areas adjacent to the Chicago River. Informal concrete sitting ledges also draw people to the river providing good

Figure 1.2   The Exterior of the McCormick Bridgehouse and Chicago River Museum Features a 1928 Bas-Relief by Henry Hering, titled "The Defenders." It commemorates the 1812 Battle of Fort Dearborn. Note the sign shown in the bottom-left corner of the photo. It indicates one of the many entrances of the Riverwalk. *Source*: Photo by the author.

vantage points for people-watching, given the nearly constant flow of pedestrians who walk along the river. Also, these sitting spaces offer a beautiful view of the memorial, the river, and the skyline, while evoking a sense of being in a more natural environment (figure 1.3; Al-Kodmany, 2017).

Phase 2

Phase 2 extends from State Street to LaSalle Street. Completed in 2015, it contains three "urban rooms," a concept that was promoted by Daniel Burnham and Edward H. Bennett in their Chicago 1909 Plan. Spanning approx. 300 feet, each "urban room" features a distinct river-inspired theme and is bordered by DuSable Bridges, which promote a sense of visual enclosure. These "urban rooms" are the Marina Plaza, the Cove, and the River Theater, as follows.

The Marina Plaza

The Marina Plaza is located between State Street and Dearborn Street. Its name was inspired by Marina City—a mixed-use skyscraper complex located

Figure 1.3 Chicago's Vietnam Veterans Memorial. The first phase of the Riverwalk proj-ect involved relocating the Vietnam War Memorial from Wacker Drive to a space closer to the river. The top picture shows the memorial at the river level, while the bottom picture shows the memorial fountain and staircase that leads to Wacker Drive. The memorial provides plenty of spaces for people to sit, relax, enjoy the plaza, and watch the river life. *Source:* Photo by the author.

across the river. This "urban room" incorporates restaurants and outdoor seating spaces. One of the notable seating options is elegantly detailed and custom high-backed teak benches. The plaza also features steps that descend nearly to the river level, providing visitors an intimate connection with water (figures 1.4 and 1.5).

The Cove

The Cove is located between Dearborn and Clark streets. It contains eatery places and diverse seating options with a landscaping theme that was inspired

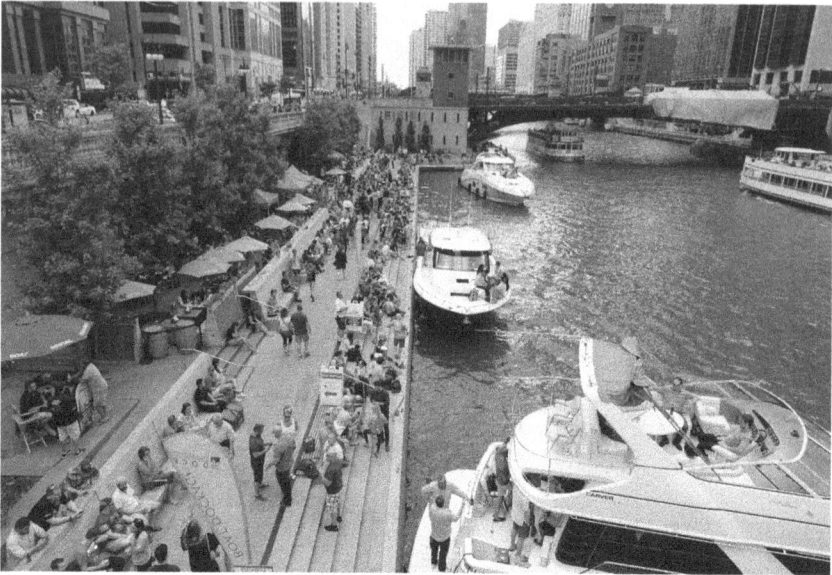

Figure 1.4   A Typical Condition of the Chicago Riverwalk before Retrofitting Efforts.
Specifically, this is the older condition of the renovated Marina Plaza. Note that the place
was characterized by abandoned arcades and narrow path. It was disinviting, deserted,
and suffered from discontinuity. Pedestrians had to use the stairway every time when they
needed to move to next section of the path. The upper part of the photograph shows a
portion of the Upper Wacker Drive. *Source*: Courtesy of the City of Chicago; https://www
.chicago.gov/city/en.html

by beach design. As such, visitors can spread on low-slung benches that are
reminiscent of beach stones, and they can also enjoy beachgrasses, wetland
plants, and trees with woodland ferns. Designed with shallow edges, it is
closer to water than other spaces to ease access to kayaks that can be docked
and rented in this place. Since the plaza is of lower elevation, its planting
beds were designed to tolerate floods when submerged by river water during
storms and heavy rain events.

The River Theater

The River Theater extends from Clark Street to LaSalle Street. Its prime feature is a sculptural staircase that contains lots of seating area and is penetrated
by a diagonal pedestrian ramp (with less than 5% gradient) that connects the
Riverwalk with the Upper Wacker Drive. Seventeen mature trees with large
canopies sprinkle the space, providing visitors with shade and adding natural
beauty (figure 1.6). As tall trees attract birds, visitors may also enjoy listening
to their twittering.

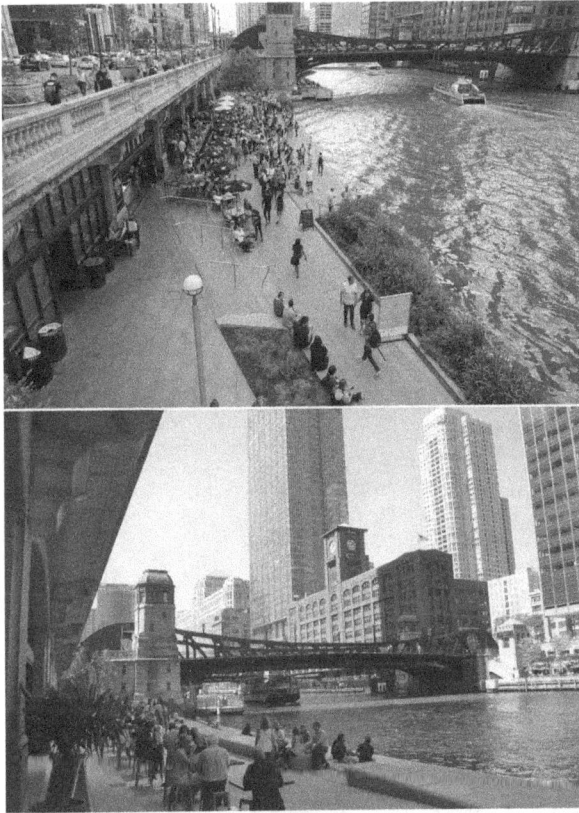

Figure 1.5   The Marina Plaza. When compared with the previous photograph (figure 1.5), we note how the retrofit project has drastically improved the space; the project transformed abandoned arcades and narrow path into a vibrant social space. The Riverwalk extends 20 ft out into the river and brings visitors closer to the river level so that they enjoy intimacy with water. The new plaza is a busy place with people socializing, walking, relaxing, eating, drinking and sitting on ledges, chairs, and boats. *Source*: Photo by the author.

Phase 3

Phase 3 extends from LaSalle Street to Lake Street. Completed in 2016, it consists of three "urban rooms," including the Water Plaza, the Jetty, and the Riverbank (previously called "Boardwalk"). Each room spans 300 feet except the Riverbank, which spans 500 feet. Similar to "urban rooms" in Phase 2, these rooms are also bordered by DuSable Bridges, enhancing the sense of visual enclosure.

The Water Plaza

The Water Plaza extends from LaSalle Street to Wells Street. It contains a long, rectangular zero-depth water fountain that features a black granite

Figure 1.6   The River Theater. The sculpted theater offers a strategic place to watch the river's lively activities (boating, kayaking, canoeing, etc.); to have a private conversation, eat lunch, photograph the river, or play with electronic gadgets. As tall trees attract birds, visitors may also enjoy listening to their twittering. The bottom photo shows a Floating Museum exhibit in August 2017 placed on a barge next to the River Theater. *Source*: Photo by the author.

pavement and interactive water jets that engage children in touching and playing with water. Acting as a splash pad, the plaza also often engages people of all ages and pets. The Water Plaza compensates for lacking the opportunity to play with the river water. It also contains public restrooms as well as mechanical equipment that runs the water jets (figure 1.7).

The Jetty

Spanning the area from Wells Street to Franklin Street, the Jetty comprises seven piers that protrude over the river, providing suitable places for people

Figure 1.7   The Water Plaza. The introduced plaza is a lovely place where children and families enjoy playing with water. *Source*: Photo by the author.

to practice fishing (though, an appropriate license is required), watch fish, and enjoy an ecological experience and social intimacy. In this area, underwater fish habitats were installed, and floating gardens that provide a healthy habitat for the fish population were integrated. Anchored by stainless steel pylons, the floating gardens can rise up to 8 feet in the case of flooding. The Jetty uses water-tolerant plants in both the floating wetlands and water gardens. In addition to ecological and recreational purposes, the floating gardens promote educational opportunities by enabling visitors to learn and interact with the aquatic ecosystem (figure 1.8).

To support subsurface aquatic habitats in this area, fish ecologists used creative intervention techniques. For example, they integrated seven limnetic "habitat curtains," where each consists of steel frames and steel wire mesh, from which nylon ropes dangle. These are made to make algae grow on them,

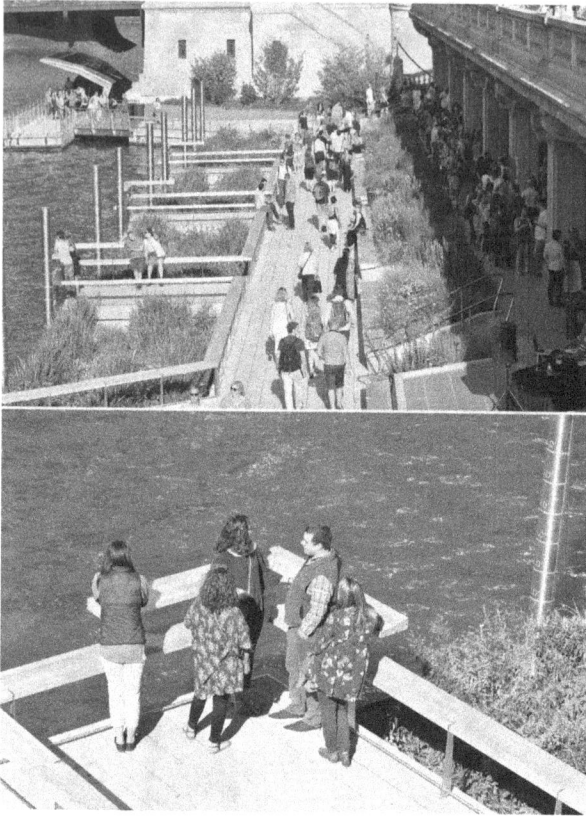

Figure 1.8   The Jetty. The integrated floating eco-gardens and rain gardens promote a unique socio-ecological learning experience (top). In addition, the Jetty offers visitors a place of solitude, privacy, and intimacy with nature (bottom). *Source*: Photo by the author.

providing food for fish. These curtains also offer an artificial filamentous substrate for inhabiting sessile organisms such as barnacles. Ecologists also wrapped around and attached "pole hulas" to the underwater structural poles, located closer to the walkway. Pole hulas are nylon ropes that promote algae growth and the breeding of amphibious insects, which are major food sources for fish. Further, they installed caisson-mounted "lunkers," which are porous steel cylinders that offer a place for fish to shelter from powerful river current and aquatic predators, larger fish, and mammals. Fish lunkers were installed on seven of the caissons that are away from the walkway. A 10-inch (25.4-centimeter) gap between the lunker and the caisson provides a shelter for fish. Interestingly, these ecological techniques that were implemented in the Jetty were partially inspired by the "fish hotel" project that attracted and provided habitat for river species before the Riverwalk was built (Hill, 2019, p. 236).

The Riverbank

The Riverbank is the last "urban room" in the Riverwalk. It extends from Franklin Street to Lake Street and contains a spacious lawn for people to sit, relax, and watch the river's activities. It also contains chairs, including Adirondack chairs that visitors use to lounge, converse, and enjoy the sunshine during their lunch hour. A pedestrian, wheelchair-accessible ramp links the Upper Wacker Drive with the Riverwalk. The city continues to explore possibilities to reconfigure this urban room (figure 1.9).

To ensure the continuity of the Riverwalk, the project created under-bridge crossings, located directly under the existing bridges (figure 1.10). As such, pedestrians can walk from Lake Street to Lakeshore in an uninterrupted manner. These "under-bridges" have canopies to protect pedestrians from dirt that may

Figure 1.9 The Riverbank. The renovated space attracts pedestrians to sit, relax, read, eat, or just watch the river's life and surrounding iconic skyscrapers. *Source*: Photo by the author.

Figure 1.10 Sheltered Under-Bridge Crossings Connect "Urban Rooms" along the Riverwalk. Note how the stainless-steel canopy beautifully reflects the river's water texture, color, and pedestrians' motions. *Source*: Photo by the author.

fall on them from the bridge deck above. They are made of stainless-steel that beautifully reflects the river's water texture, color, and pedestrians' motions. In addition, "under-bridges" create natural breaks among the various "urban rooms."

The Riverwalk also integrates LED lighting systems that promote safety and invite visitors at night. Lightings also help to define edges of the various spaces such as seating areas, gathering spaces, and the diagonal ramp of the River Theater. The integrated lighting systems can endure occasional flooding. Planting and paving systems were also designed to be resilient to expected inundation.

EVALUATING THE PROJECT: A SUSTAINABILITY FRAMEWORK

Chicago's sustainability journey continues forcefully. The city aims to be among the most sustainable cities on the planet. It viewed the Riverwalk project as a vehicle to achieve this overarching goal. Therefore, it would be fitting to evaluate the Riverwalk project within a sustainability framework. This is illustrated by examining the project against the sustainability's three dimensions, namely social, economic, and environmental, as follows.

Social Dimension

Connectivity and Accessibility

As mentioned earlier, among the prime goals of constructing the Riverwalk were to create an uninterrupted, car-free promenade that fosters enjoyable pedestrians' environment and supports intermodal connections among bike-share stations and public mass transit nodes (bus, train, and boat). These goals are certainly accomplished. As a linear urban park, the Riverwalk provides continuity and connectivity for people by knitting together various parts of the Chicago's central business district. It enables, for the first time, pedestrians to walk freely, uninterrupted by automobiles or physical obstacles, from Lakefront to the river confluence passing through the heart of Chicago downtown. This east-west circulation system is also reinforced by the south-north circulation system provided by streets that lead to the Riverwalk. Examples include Franklin Street, Wells Street, LaSalle Street, Clark Street, Dearborn Street, State Street, Wabash Street, and Michigan Avenue. The path also helps hundreds of thousands of workers in the Chicago Loop to get around. The path also provides vertical connections, mainly between Wacker Drive Viaduct and the river level.

Therefore, the Riverwalk links multiple critical areas, amenities, and facilities in downtown. People use the path to connect with the Magnificent Mile, Millennium Park, Maggie Daley Park, Navy Pier, and office spaces and commercial businesses. The path is accessible year-round. Admirably, the path adheres to the ADA requirements. As such, it is accessible by people with limited physical mobility. By integrating ramps and improving elevators, the path establishes universal access between streets and the river level. This is particularly appreciated when knowing that the path was not accessible by people with disabilities. Importantly, the 1.25-mile (2-kilometer) trail is conveniently walkable for being relatively short. This is also appreciated when compared to the 18.5-mile (30-kilometer) linear park along Lake Michigan. It is also considerably shorter than the famed 15-mile (24-kilometer) San Antonio River Walk.

Furthermore, the Riverwalk reinforces intermodal connections. It facilitates stronger links among various means of transportation, including automobiles, boats, bikes, buses, and trains. Several train stations of the CTA and METRA are close to the Riverwalk. Of course, it also connects people with water taxi and boat docks along the river. That is, people use the Riverwalk daily to walk to their commute points and stations. In addition, the path provides access to loading and storage spaces that are needed to support the river's business operations. In a nutshell, uninterruptedness of the path, unity of design, landscaping themes, diversity of activities of plazas, closeness to nature, ease of access, intermodal connectivity, walkable distances, seating places, eatery places, programmed events, and stunning skyscrapers that flank the river, collectively create a vibrant social and visually interesting environment.

Diversity of Activities

The design of the path introduces creative land-water typologies, seen in its "urban rooms," such as Marina Plaza, Cove, River Theater, Water Plaza, Jetty, and Riverbank. The design and spatial layout vary so that each room has a distinct identity. In addition, each space offers different functionalities and experiences. Further, the McCormick Bridgehouse & Chicago River Museum and the Vietnam Veterans Memorial serve educational and cultural purposes. Overall, along the path people engage with various social activities such as walking, chatting, relaxing, jogging, running, eating, dining, drinking, photographing, sightseeing, and fishing. Also, some visitors play yoga and workout. Other people could be seen biking (or walking the bikes), scooting, hoverboarding, or Segwaying. On the river, visitors can enjoy boating, kayaking, canoeing, cruising, and joining architectural and recreational tours. One can also observe various recreational boat types, like yachts, personal boats, and powerboats.

The concept of diversity is further stressed by providing distinct public art and cultural programming. Among the outstanding large-scale public art is Art on the MART. By covering 2.5 acres (1 hectare) of the MART's river-facing façade, it is one of the world's largest video-projection art installation. In 2018, the art show attracted 32,000 attendees. Further, the Floating Museum in 2017 garnered a great interest with over 90,000 attendees (Hanson and Callone, 2019). In addition, temporarily public art often sprinkles the path. As such, the Riverwalk promotes social diversity, visual interest, and joy.

Significantly, retrofitting the river's bank has enabled hundreds and thousands of visitors, whether on land or water, to enjoy excellent Chicago architecture, exhibited by iconic skyscrapers. For example, a few skyscrapers that sit side by side along the bend of the Main Branch (Trump International Hotel & Tower, 330 North Wabash, and Marina City) create a remarkable edge. Interestingly, each skyscraper represents a major architectural style and authentic design philosophy. Although these skyscrapers feature different color schemes, textures, and heights; they get along and stand proudly telling an epic story of the city's architecture. The outstanding location, significant height, and excellent architecture of Adrian Smith's Trump Tower, the city's second-tallest building, make this tower an unmistakable landmark.

Likewise, the dark exterior of Ludwig Mies van der Rohe's 330 North Wabash makes the tower prominent and establishes a pleasant contrast with the lighter colors of the Trump Tower. Similarly, Bertrand Goldberg's Marina

City features an outstanding organic profile that evokes a splendid contrast with Mies's 330 North Wabash. Further, the clustering of iconic skyscrapers creates an attractive skyline, highly appreciated as seen from close distances along the Riverwalk. The skyline night views are particularly attractive. The outcome is "a powerfully memorable canyon of space, lined by skyscrapers and punctuated by Chicago's iconic bascule bridges" (Agency, Landscape & Planning).

Further, aesthetically, visitors can enjoy and appreciate artful design of the Riverwalk. Its upper part is designed with a refined stone that matches the limestone of the Wacker Drive Viaduct; seen in the Beaux-Arts retaining wall, stairways, balustrade, lighting posts, and bridge houses. As such, the new design reinforces the historic integrity of the area. In contrast to refined and formal design at the upper level, at the river level (a flood-prone area), the design uses rugged precast concrete and metal structures. Overall, the design balances the formality of the Beaux-Arts design with the informality of the river's plaza and "nakedness" of the striking steel superstructure Bascule Bridges.

Inclusivity

Notably, accessing the river's social, cultural, and educational activities and gathering places are largely free of charge. Consequently, the Riverwalk attracts a broad spectrum of people, including residents, local and global visitors, business owners, professionals, and public officials. Free public spaces promote inclusivity, and the mixed crowd that the Riverwalk attracts reflects Chicago's cultural values of equality and diversity. Among the most beautiful feelings that visitors experience is sharing space and experiences with people from all walks of life. Their combined presence creates a vibrant downtown, engenders strong social bonds and an enduring sense of belonging.

The value of maintaining public amenities is well established in Chicago. Beginning with the 1909 Plan and until today, free public access along the entire lakefront to significant public places such as Millennium Park, Maggie Daley Park, Chicago Riverwalk, and Navy Pier is provided year-round. This quality is also observed in city-owned public plazas (e.g., Federal Plaza and Daily Plaza) and privately owned public plazas (e.g., Chase Plaza and 311 South Walker Plaza). Free admission is also provided for seasonal festivals in events such as Blues in June, Taste of Chicago in July, Jazz over Labor Day, and World Music in September.

Walking Benefits

The Riverwalk also encourages people to engage into a healthy habit—walking! Walkable cities are increasingly desirable. Overall, emphasis on walking prevails for numerous reasons, such as health benefits, social inter-action, environmental well-being, and energy savings, to name a few. Dan Burden, an expert on walkability, makes a good point, "Walkability is a word that did not exist just 20 years ago. We made walking so unnatural that we had to invent a word to describe what we were missing . . . Essentially, walkability is allowing people to do what the human body was designed to do in the first place: to go places without having to get into some mechanical instrument" (Burden, 2014). Walking is the simplest form of exercise; people are pedestrians by design. In his book *Walkable City: How Downtown Can Save America One Step at a Time*, urbanist Jeff Speck (2013, p. 38) poetically explains the importance of walking by stating: "As a fish needs to swim, a bird to fly, a deer to run, we need to walk, not in order to survive, but to be happy." A daily walking routine of 20 minutes can prevent heart disease, diabetes, depression, and some cancers (Elliott, 2010).

Walking increases endorphin production and neuron development, and it can also lower blood pressure, ease back pain, strengthen arms and legs muscles, improve joint conditions and balance, and reduce the risk of glau-coma and osteoporosis. In addition to health benefits, walking can promote a city's resilience and cultural heritage, reduce crime, foster creative thinking, enhance productivity, improve placemaking, and increase land and property values. Jeff Speck (2013, pp. 45–46) argues that walkability is a critical fac-tor to thriving cities. He insists that sustainable places should free us from dependence on the automobile, which he calls "a gas-belching, time-wasting, life-threatening prosthetic device." He lamented that suburban sprawl is the worst urban planning model we produced, characterized by "fattened roads, emaciated sidewalks, deleted trees, fry-pit drive-thrus, and 10-acre parking lots."

Economic Dimension

The Riverwalk has opened up an entirely new entertainment district for locals and tourists alike. It has invited venders and retailers and increased sales. According to research by Hanson and Callone (2019), since opening of the second and third phases of the Riverwalk, the number of vendors has dou-bled, and their businesses have been improved substantially. A larger number of visitors has increased parking revenue. Similarly, collected fees from boat parking and vendors have increased. Tourists have amplified hotel profits. Overall, these activities supported existing jobs and created new ones. From 2016 to 2019, the project has created additional 66 permanent, 170 seasonal, and 125 part-time jobs (Landscape Architecture Foundation). Of course, in addition, the project created significant construction jobs.

The Riverwalk project has increased property value and the city's property tax base. It has enhanced the marketability of Wacker Drive as a world-class business downtown and improved the opportunity to attract top employees and companies. According to Sasaki, the project has stimulated over $8 billion in economic developments. Also, hotel tax, retail sales tax, local business tax, and income tax have increased. In just six years (2013–2018), the project has gathered over $16 million toward the repayment of the $99 million federal loan that was granted to construct the second and third phases of the project. The loan repayment methods included fees and revenues from tour boats, private boats, boat docking, motor fuel tax, outdoor advertising, sponsorship income, rent, and leases (Landscape Architecture Foundation).

The costs of the project are relatively reasonable. They are estimated in hundreds of million dollars, whereas each of the surrounding skyscrapers could cost as much or even more. For example, the cost of constructing the ninety-eight-story Trump Tower was $847 million; the cost of the fifty-two-story River Point was $500 million; and the cost of the fifty-four-story 150 North Riverside was about $500 million (CTBUH Skyscraper Center). In this regard, Gina Ford, design principal of Sasaki, commented that architectural aesthetic is certainly important. However, with limited costs, improving public spaces can make a significant difference to masses of people. Sasaki also conducted a survey asking 1,000 urban dwellers about their urban experiences and 67% responded that their best moments were in civic spaces and public parks (Sisson, 2016). Collectively, the Riverwalk project presents a case where retrofitting an aging infrastructure is economically worthwhile.

Environmental Dimension

One of the important goals of the Riverwalk project was improving the environmental well-being of the river. So far, the Riverwalk is promoting the river's recovery and restoration of on-land and in-stream habitats. As a result of reintroducing native trees, plants, and grasses, water quality has been enhanced—vegetation's roots absorb and break down pollutants. Cleaner water has started bringing back local fish, birds, turtles, and muskrats. Ergo, this process is meant to pave the way to establish a holistic ecosystem that invites land aquatic animals, who, through their natural habits, also help in removing toxins from the river. Several bird species have been already invited; examples include the mallards, ring-billed gull, house sparrow, herring gulls, rock pigeons, American robins, European starlings, and Peregrine falcons. There are also proposals to integrate underwater plant species to attract marine life (Sisson, 2016). Aesthetics are further enhanced by having greeneries cover the unsightly river's steel walls.

The floating eco-gardens, or floating wetland islands, reintroduce plant species native to Illinois wetlands and prairies to support multiple ecological functions, for example, cleansing the river's polluted water, mitigating floods, and attracting and sustaining wildlife. Plants' roots do not reach the riverbed because it is much deeper. As such, plants' roots have intimate, symbiotic interaction with the river water, increasing their function of cleansing the water, while water constantly nourishes plants. Besides, plants reduce nitrogen and phosphorus levels in the river. Consequently, the fish population has already been increasing around the floating eco-gardens, which in turn has been attracting land-water animals such as turtles. According to FOCR, forty years ago, only seven aquatic species were present in the Chicago River, while today, more than seventy-five species live there.

Therefore, the Riverwalk project contributed to sensible ecological improvement in the river. According to research by Hanson and Callone (2019), the Floristic Quality Index (FQI) has increased from 0 to 38.2 in an area of 19,529 sq. feet (1,814 sq. meter). Further, according to FOCR, forty years ago, only seven aquatic species were present in the Chicago River, while today, more than seventy-five species live there. In addition to serving ecological functions, the floating gardens serve as an outdoor classroom for learning about fish, fauna, and critters. Inviting people to the river makes them think about the importance of their natural and ecological functions. Public programming along the Riverwalk also promotes environmental awareness of climate change, flooding resiliency, and sustainable design. For example, "the 2017 monthly Fish Parades" and "2018 children's environmental" programs have offered children opportunities to fish and observe thousands of fish, representing dozens of species (Hanson and Callone, 2019; Hsieh et al., 2018).

In addition to ecological features in specific areas, the Riverwalk employs sustainable and ecological measures throughout by using and integrating:

- continuous planting trenches that allow trees and shrubs to share a larger soil volume;
- native plants, such as river birches and ephemeral plants that are tolerant to changing climatic conditions;
- resilient, inundation-tolerant plants near the river, most vulnerable area for flooding;
- rainwater collectors for irrigation to save on portable water;
- LED lighting for pedestrians to promote safety and invite visitors at night;
- LED "fish light" that runs along the entire jetty's edge, illuminating the fish habitat beneath the water;
- recycled materials such as reclaimed teak;
- materials that can be cleaned of debris by power washing after a flood;
- paving that incorporates stormwater-recycling irrigation system;

- paving that accommodates large volumes of soil to sustain the health of planted trees; and
- educational programs to educate about ecological restoration.

However, the journey of improving the Chicago River conditions is not over. Although the Riverwalk project has helped in improving water quality and attracting fish species (e.g., largemouth and smallmouth bass, bluegill bream, crappie, and sunfish), the ecological performance still needs improvement. For example, the river's water continues to be polluted, causing health problems to plants and aquatic habitats. Also, unclean water makes canoers and kayakers less comfortable. Visitors are advised not to even touch the river water. Beside water quality issues, vertical walls that stand along the river's banks prevent them from fostering healthy aqua habitats and improving the amount of macrophyte, overhanging vegetation. Furthermore, ecological testing suggests that sediments are highly contaminated. Their toxicity and metal threshold are excessive and, therefore, the poor quality of sediments could pose an ecological risk and prevent fish from reproducing. City employees collect dead fish around the river's main branch daily (Hsieh et al., 2018). Overall, there is a continuous need to clean-up the river and remove contaminated sediments.

EXTRAORDINARY CHALLENGES

Achieving the intended goals of the Riverwalk project is further appreciated and emphasized when we learn about the extraordinary challenges experienced in all design, implementation, and construction stages. For example, the construction had to take into account the numerous utilities and infrastructure that already exist in and near the construction site. In addition, the transport operation of the river had to continue during the construction process, imposing logistic challenges. For example, the river was used to deliver all construction materials to the site via barges. The U.S. Coast Guard, however, does not allow to occupy more than half the width of the river at any moment for any purposes, including delivering materials and construction activities, to evade obstructing regular river traffic. With a limited work area (20–25 feet/6–7.6 meters), the delivery and storage of materials as well as construction process had to be carefully planned. Making things more complicated, river traffic (e.g., commercial ships, tourist boats, personal watercrafts, canoers, and kayakers) increases substantially during summer time, a prime time for construction activities.

Building the pedestrian connections in a confined space beneath the six Bascule Bridges was extra challenging. The process required opening the

Bascule Bridges and using giant cranes, placed on barges, to anchor pillars and slabs. Building caissons for these pillars preceded the insertion. Due to the swampy soil, each caisson had to extend 60 feet (18.2 meters) down to reach the riverbed below. Since opening bridges interrupts auto and pedestrian traffic on major streets in the busy downtown, the construction process had to be completed in the shortest time possible. However, marine engineering, by default, requires more time, making the construction process extra challenging.

Overall, the project required technical ingenuity and advanced marine construction methods. For example, the construction had to accommodate the river's annual flood dynamics and fluctuations that measure up to 7 feet vertically. Similarly, on the vertical plane, the design had to account for the varying heights of several key elements, including Upper Wacker (+22 feet/+0.7 meter), Lower Wacker (+5 feet/+1.5 meter), and the River (-2 feet/-0.6 meter, on average). Also, it was challenging to meet the ADA (Americans with Disabilities Act) requirements, ensuring that the path is accessible by people with limited physical mobility. Overall, the construction scheduling had to be "very linear;" that is, it was difficult to work on multiple construction tasks simultaneously. The tight scheduling and working in confined and complex spaces that contain utility lines mandated the process to be error free. Of course, construction projects in water engenders additional work of sealing and waterproofing all elements including joints, for example, wall joints and paver joints. Collectively, the design, implementation, and construction of the Riverwalk were exceedingly challenging.

Deservedly, the project has received immense praise, recognition, and awards. For example, in honoring Sasaki's consulting firm with the 2018 ASLA Award, the jury stated that the Riverwalk project represents an extraordinary collaborative work among interdisciplinary teams of architects, landscape architects, urban designers, community organizations, government agencies, and public officials, among others. After long years of earnest commitment, the outcome is an additional jewel that connects Chicago Downtown with the lakefront. Indeed, the project demonstrates novel engineering and innovative technical approaches in an exceedingly complex urban setting. It solves challenging problems while observing sustainability. The project represents a creative retrofit to a neglected urban infrastructure; it is a welcome addition to Chicago's growing body of outstanding parks.

CONCLUSIONS

The Chicago Riverwalk project has transformed an underutilized, defunct, and aging urban infrastructure into a downtown amenity and a tourist attraction. For almost a century, the Chicago River was heavily polluted to the extent that it deterred social life, economic activities, urban development, and damaged the river's ecosystem. The Riverwalk project has improved the river's conditions, partially restored its ecosystem, and made it a comfortable public destination. After serious work and genuine collaboration, the Riverwalk has become very much the heart and soul of Chicago. It is so loved by its residents, national and international tourists alike. Importantly, the Riverwalk lives up to Burnham's vision for a civic circulation path that connects different areas and creates an entirely new way to get around the city on foot and boats. Sasaki described the project as a sophisticated urban retrofit that even Burnham himself never imagined.

Many cities around the globe suffer from polluted and aging river infrastructure. As such, the Riverwalk project may present an exemplary case study. One important lesson that we may learn is that revamping aging urban infrastructure requires sustained hard work, serious collaboration, and an earnest commitment by local, state, federal governments as well as community-based organizations, individuals, businesses, and institutions. From vision to completion, the project took over twenty-five years. Certainly, there were moments of frustration and give ups. However, persistence and perseverance have helped to keep the momentum of this project. Overall, the Chicago Riverwalk project presents a useful case study that may inspire, inform, and benefit many cities around the globe. This notion could be emphasized further since many people, such as youngsters, the "creative class," and elderlies are returning to the city due to favoring urban environments with cultural amenities and public transport. Finally, innovative projects are likely to attract international tourists, thereby improving economy. It is hoped that this chapter updates and stimulates other cities in order to promote a livable and sustainable world.

ACKNOWLEDGMENTS

The author would like to thank anonymous reviewers for useful feedback. Also, sincere thanks go to Dr. David S. K. Ting for encouragement.

REFERENCES

2018 ASLA Professional Awards. A Program of the ASLA Fund. https://www.asla .org/2018awards/453251-Chicago_Riverwalk.html. Accessed 15 February 2020.
Agency, Landscape & Planning. https://agencylp.com/projects/the-chicago-riverwal k/. Accessed 15 February 2020.

Al-Kodmany, K. (2017). *Understanding Tall Buildings: A Theory of Placemaking.* Routledge, London.

Al-Kodmany, K. (2018). *The Vertical City: A Sustainable Development Model.* WIT Press, Southampton, UK.

Bosch, J. (2008). *A View from the River: The Chicago Architecture Foundation River Cruise.* Pomegranate Communications, Portland, OR.

Burden, D. (2014). The Power of Walkability. *Blue Zones,* November 18. https://ww w.bluezones.com/2014/11/power-walkability/. Accessed 5 December 2019.

Chesla, E. (2017). Investing in Urban Infrastructure, The 2017 Rudy Bruner Award for Urban Excellence. http://www.rudybruneraward.org/wp-content/uploads/2017/ 01/05-Chicago-Riverwalk-Phases-2-3.pdf. Accessed 15 February 2020.

Chicago. 7 Things You Didn't Know About the Chicago Riverwalk. https://ww w.choosechicago.com/articles/parks-outdoors/7-things-you-didnt-know-about-the -chicago-riverwalk/. Accessed 15 February 2020.

Chicago Architecture Center. McCormick Bridgehouse & Chicago River Museum. https://openhousechicago.org/sites/site/mccormick-bridgehouse-chicago-river -museum/. Accessed 15 February 2020.

CTBUH Skyscraper Center. https://www.skyscrapercenter.com. Accessed 15 February 2020.

Dyja, T. L. (2014). *The Third Coast: When Chicago Built the American Dream.* Penguin, NYC.

Elliott, D. (2010). A Useful Tool with Room for Improvement, Planning, the Magazine of the American Planning Association, December, pp. 38–43.

Friends of the Chicago River. 40 Years, 40 Key Moments. https://www.chicagoriver.org/blog/2019/2/friends-of-the-chicago-river-40-years-40 -key-moments. Accessed 15 February 2020.

Godschalk, D., & Malizia, E. (2014). Sustainable Development Metrics. *Planning, the Magazine of the American Planning Association,* February, pp. 22–26.

Godschalk, D. R., & Rouse, D. C. (2015). *Sustaining Places: Best Practices for Comprehensive Plans,* PAS Report 578. https://www.planning.org/publications/re port/9026901/. Accessed 5 December 2019.

Hanson, S., & Callone, M. (2019). Chicago Riverwalk, Phases 2 & 3 Methods. *Landscape Performance Series.* Landscape Architecture Foundation. https://doi .org/10.31353/cs1501. Accessed 15 February 2020.

Hellenthal, M. F., & Gross, D. M. (2016). Design and Construction of the Chicago River. *Geo-Chicago.* https://doi.org/10.1061/9780784480120.065.

Henchen, M., & Kroh, K. (2020). *A New Approach to America's Rapidly Aging Gas Infrastructure.* Rocky Mountain Institute. https://rmi.org/a-new-approach-to-a mericas-rapidly-aging-gas-infrastructure/. Accessed 15 February 2020.

Hill, L. (2019). *The Chicago River: A Natural and Unnatural History.* Southern Illinois University Press, Carbondale, IL.

Hsieh, H., Li, X., Wang, S., & Wu, Y. (2018). *Post-Occupancy Evaluation of the Chicago Riverwalk,*

Final Report Submitted to Sasaki. https://deepblue.lib.umich.edu/handle/2027.42 /143161.

Hunt, B., & DeVries, J. B. (2017). *Planning Chicago*. Routledge, London.

Johnson, E. W. (2006). *Chicago Metropolis 2020: The Chicago Plan for the Twenty-First Century*. The University of Chicago Press Chicago, Chicago, IL.

Landscape Architecture Foundation. Chicago River Phases 2&3. *Landscape Performance Series*. https://www.landscapeperformance.org/case-study-briefs/chicago-riverwalk. Accessed 15 February 2020.

Office of the Mayor. (2012). https://www.chicago.gov/city/en/depts/mayor/press_room/press_releases/2012/february_2012/mayor_emanuel_announcesnewsustainability-focusedtwitterfeedandwe.html. Accessed 15 February 2020.

Sasaki. (2019). The Dynamism of the Chicago River, October 24. https://www.sasaki.com/voices/the-dynamism-of-the-chicago-river/. Accessed 15 February 2020.

Sisson, P. (2016). Chicago's New Riverwalk Offers a Vision of the Future of Urban Parks. *Curbed*. https://www.curbed.com/2016/10/24/13382868/chicago-riverwalk-landscape-architect-urbanism-design. Accessed 15 February 2020.

Smith, C. S. (2006). *The Plan of Chicago: Daniel Burnham and the Remaking of the American City*. The University of Chicago Press, Chicago, IL.

Speck, J. (2013a). *Walkable City: How Downtown Can Save America One Step at a Time*. North Point Press, NYC.

Speck, J. (2013b). The Walkable City. *TED^x Ideas Worth Spreading*, October. https://www.ted.com/talks/jeff_speck_the_walkable_city. Accessed 5 December 2019.

Speros, W. (2016). The Chicago Riverwalk Opens to the Public. *Contract*, November 2. https://www.contractdesign.com/news/projects/the-chicago-riverwalk-opens-to-the-public/. Accessed 15 February 2020.

Sustainable Chicago. (2015). https://www.chicago.gov/content/dam/city/progs/env/SustainableChicago2015.pdf. Accessed 15 February 2020.

United Nation, HABITAT, World Urban Forum (WUF). (2017). https://wuf.unhabitat.org/node/145. Accessed 5 December 2017.

World Commission on Environment and Development (WCED). (1987). *Our Common Future*. Oxford University Press, Oxford, UK.

Chapter 2

Sustaining Old Infrastructures with Advanced Window Retrofitting Technologies

Rui Kou and Ying Zhong

INTRODUCTION

In the United States, there are over 85 million existing residential and commercial buildings consuming ~40% of the total energy consumption of the entire country (DOE, 2014), while ~40% (~15 quadrillion British thermal units, shorted as "quads") of which is consumed by building heating ventilation and air conditioning (HVAC) (U.S. Department of Energy, 2014; Afram and Janabi-Sharifi, 2014; Lu, 2012). Windows, as one of the most essential and costly elements of the building envelopes, leak about 4 quads of energy per year because of their poor energy insulating property (Arasteh et al., 1998; Abonyi et al., 1999; DOE, 2014). Figure 2.1 shows the estimated primary energy required to support the energy consumption by the windows in the four broad Census Bureau regions in the United States ("Single-Pane Highly Insulating Efficient Lucid Designs (SHIELD) Program Overview" 2014) In higher-cost modern buildings, double-pane or even triple-pane windows have been adopted for desirable thermal insulation properties. Up until now, 30–40% of the existing windows are still single-pane windows that are responsible for over 50% of the total energy leakage through the widows (DoE, 2011; Klems, 2003). Upgrading these windows in existing buildings has been extremely slow due to the high cost of window replacement (about $50~100/ sq. ft (DoE, 2011)), incompatibilities of size and weight of double-pane windows compared with the original single-pane units, and undesirable reduction in visible transmittance (VT) and change in color rendering index (CRI) of double-pane replacements. Only 2% of the existing single-pane is replaced per year ("Single-Pane Highly Insulating Efficient Lucid Designs (SHIELD) Program Overview" 2014). On the contrary, retrofitting windows is a more cost-effective way which can potentially return $12 billion per year to energy consumers ("Single-Pane

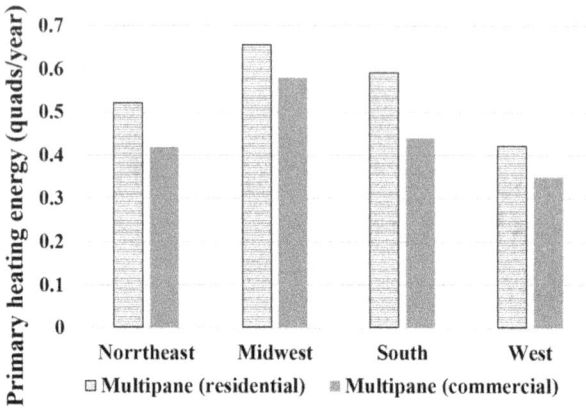

Figure 2.1 Estimated Primary Energy Usage for Heat Flow through the Windows of Existing Buildings in Four U.S. Census Regions. The results include the energy usage for all of single-pane, double-pane, and multi-pane windows in residential and commercial buildings. The area fraction of the single-pane windows in each region is shown at the bottom of each bar. *Source*: Created by the authors.

Highly Insulating Efficient Lucid Designs (SHIELD) Program Overview" 2014). If the total remaining stock of single-pane windows can be successfully retrofitted, it will reduce total energy consumption by 1.2 quads per year (DOE, 2014). Reducing the energy lost through the single-pane windows will not only directly impact both residential homes and commercial business by saving money for homeowners and business owners, but it will also mitigate environmental impacts of the energy lost (Horowitz et al. 2011; Carmody et al. 2004). Retrofitting single-pane window with highly transparent thermal insulating materials becomes more and more important.

HEAT TRANSFER MECHANISM AND U-FACTOR

Heat transfers occur from the room side to the exterior side through window-panes by conduction, convection, and radiation, as shown in figure 2.2. All these three methods of heat transfer should be taken into consideration to mitigate the heat loss through windows.

Heat Conduction

Conduction is a type of heat transfer that occurs within a solid, liquid, or gaseous medium, for example, window glass, frame, and air. The microscopic colliding particles and moving electrons within a solid material result in this internal energy transfer, as shown in figure 2.3.

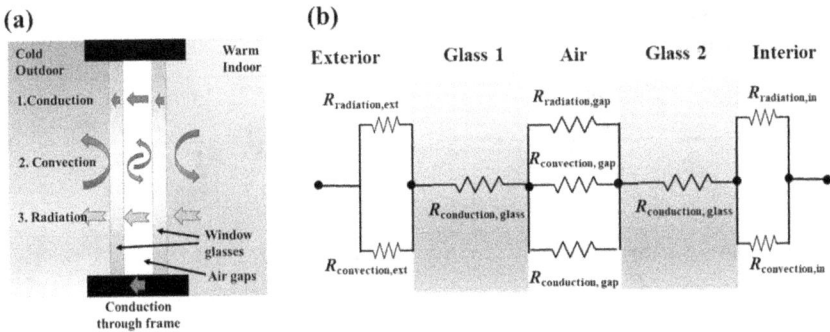

Figure 2.2 (a) The Schematic of Three Main Heat Loss Mechanisms through Windowpane. (b) The Thermal Resistance Network of Double-Pane Window. *Source:* Created by the authors.

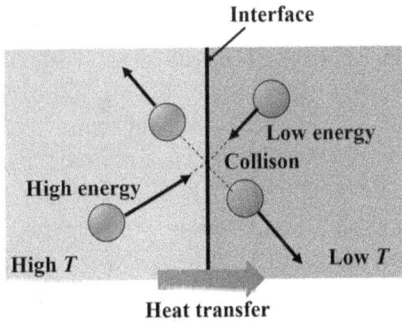

Figure 2.3 Heat Conduction: Molecule Collisions Occur at the Interface (Contact Surface) Exchange the Energy from the High Temperature to the Lower Region. The collision increases the energy of the particle with lower energy and decreases the energy of the higher one, which results in the heat transfer from the high-temperature end to the low-temperature end. *Source:* Created by the authors.

For the single-pane window, when there is a temperature difference between the room side and exterior side, the conductive heat transfer rate through window glass and its frame follows the Fourier law (Bergman et al., 2011):

$$q_{conduction} = -\kappa_{medium} \frac{\Delta T}{t} \tag{2.1}$$

where, ΔT represents temperature difference across the medium, κ_{medium} is the thermal conductivity of the medium, and t is its thickness. The medium can be glass, air, window frame, etc. The thermal conductivities of some common substances related to window application are shown in Table 2.1. Metals and glass materials have relatively high thermal conductivities, while thermal conductivity of gas is much lower. The conductive heat transfer coefficient

Table 2.1 Thermal Conductivities of Common Substances Relative to Windows (room temperature data)

Substance	Thermal Conductivity κ (W/m·K)	Substance	Thermal Conductivity κ (W/m·K)
Aluminum	220	Wood	0.08–0.16
Steel (stainless)	14	Glass (average)	0.84
Air	0.026		

Source: Griffith and Arasteh (1992).

is dependent on the thermal conductivity of the medium. Replacing the metal frame with a wood frame, and especially, equipping window glass with highly thermal insulating materials with low thermal conductivities can hugely reduce the conductive heat loss through windows.

Heat Convection

Convection is heat transfer by the large-scale movement of air, which occurs in the exterior surface, interior surface of window glass, and the air gaps in multi-paned windows. Heated regions of the air will become less dense than cooler ones, and heated air tends to rise. Subsequently, cooler air will take its place. Once the cooler air is warmed again, it will rise and make the heat transfer resume. As shown in figure 2.2, during winter, the room is warmer than cold weather outside. The window section acts like a heat transfer medium from warm indoor to cold outdoor. The convective heat transfer rate in air gaps of multipaned window and at the outermost surfaces of window glass can be expressed by (Incropera et al., 2007; Wright, 1996b):

$$q_{\text{convection}} = -h_{\text{convection}} \Delta T \qquad (2.2)$$

where, h_{conv} is the convective heat transfer coefficient of air.

Heat Radiation

Radiation is the heat transfer through electromagnetic waves instead of through medium. The capacity of the material to absorb and reradiate energy is called emissivity (Schaefer et al., 1997). In winters, when the heat energy is transferred from the room to the window glass, the glass reflects some part of the energy and absorbs the rest of it. The less heat the window absorbs, the better insulating function it can provide. The radiative heat transfer rate from surface 1 to surface 2 can be expressed as (Howell et al., 2015):

$$q_{\text{radiation}} = \frac{-\sigma \left(T_1^4 - T_1^4 \right)}{\dfrac{1}{\varepsilon_1} + \dfrac{1}{\varepsilon_2} - 1} \qquad (2.3)$$

where, the Stefan Boltzmann constant $\sigma = 5.67 \times 10^{-8}$ W / (m$^2 \cdot$ K^4) and ε_1 and ε_2 are the surface emissivities of the surface 1 and surface 2. Clearly, by decreasing the surface emissivity of window glass, the radiative heat transfer can be mitigated.

U-Factor and Window Retrofitting

The rate of heat loss through the combined effect of conduction, radiation, and convection is indicated by the term U-factor of window assembly. U-factor is the inverse of R-value, which is the resistance to heat flow of windows:

$$U = R^{-1} = \frac{1}{R_{\text{external}} + \dfrac{d_1}{\lambda_1} + \dfrac{d_2}{\lambda_2} + \cdots + R_{\text{internal}}} \tag{2.4}$$

where, R_{external} and R_{internal} represent the external and internal heat transfer resistance. d_i represent the layer thickness and λ_i is the medium thermal conductivity. U-factor measures the rate of heat transfer. Higher U-factor indicates faster heat transfer and lower U-factor indicates slower heat transfer and better insulating property and less energy leakage. The U-factor values for the center-of-glazing area, edge, and frame are different, depending on the component's material and structure. Among which, the glazing area takes the major part of the energy loss through windows. Adding an insulating layer on the window glazing can significantly reduce the U-factor of windows. The U-factor of the single-pane window is around 5.8 W/m$^2\cdot$K (Apte and Arasteh, 2008) which is inefficient for thermal insulation. By retrofitting the single-pane window with a low U-factor coating (say, 2.8 W/m$^2\cdot$K), if heat energy cost is 0.1040 $/kWh, the coating can save 1391 $/m^2 in 10 years (Kou et al., 2019). Thus, retrofitting single-pane windows is much faster, cheaper, and less labor-intensive than upgrading single-pane windows. Desirably, the retrofitted window has low emissivity (low-e), low overall U-factor (<0.5), long service life (>10 years), high visual transmittance (VT) (>70%) and low haze (<1%), and high color rendering index (>0.9).

ENERGY STAR WINDOWS

Currently, most popular energy-efficient windows (e.g., Energy Star Windows) are double-pane or triple-pane windows with low-e glasses. For energy star windows, the U-factor of entire window should be less than 1.7 W/m²·K, which meet or exceed energy efficiency guidelines set by the U.S. Environmental Protection Agency ("Energy Star Performance Criteria for Windows, Doors, and Skylights" 2015). To mitigate conductive heat transfer, air or Argon (Ar) gas, etc., are always filled in the gaps between glass panes. The gap thickness ranges typically from 8 to 20 mm. Thanks to the low thermal conductivity of air (0.026 $W\cdot m^{-1}\cdot K^{-1}$), Ar (0.0162 $W\cdot m^{-1}\cdot K^{-1}$) and others, as shown in table 2.2, the overall thermal conductivity of the window section is much reduced. If the gap between glass panes is thicker than ~8 mm, the convective heat transfer cannot be ignored (Wright, 1996b). Once the gap thickness is over ~20 mm, with the increasing of the gap thickness, the convective heat transfer coefficient will be more pronounced compared to that of conduction and also lead to a significant reduction in U-factor (Wright, 1996b).

Table 2.3 shows some typical glazing units and their U-factors. DGU, TGU, and low-e indicates double glazing unit, triple glazing unit, and low-emissivity coating coated clear glass, respectively.

Table 2.2   Some Common Filled-in Gases in Multipaned Windows

Substance	Thermal Conductivity κ (W/m·K)	Substance	Thermal Conductivity κ (W/m·K)
Air	0.026	Nitrogen	0.0255
Argon	0.0162	Carbon dioxide	0.0162
$CFCl_3$	0.0083	N_2O	0.0162
CF_4	0.0160	SF_6	0.0140
Chlorofluorocarbon	0.0084	Hydrofluorocarbon	0.0140
Hydrochlorofluorocarbon	0.0110		

Source: Griffith and Arasteh (1992).

Table 2.3   U-Factors of Some Common Multiple-Pane Glazing Units

Glazing Type	W/m·K
DGU [Clear(4 mm)/Air(16 mm)/low-e(4 mm)]	1.40049
DGU [Clear(4 mm)/Argon gas(16 mm)/low-e(4 mm)]	1.20204
TGU [Clear(4 mm)/Air(16 mm)/low-e(4 mm)/Argon gas(16 mm)/low-E(4 mm)]	0.79947
TGU [Clear(4 mm)/Argon gas(16 mm)/low-e(4 mm)/Argon gas(16 mm)/low-e(4 mm)]]	0.60102
DGU [Clear(4 mm)/Argon gas(16 mm)/low-e(4 mm)]	1.29843
DGU [Clear(6 mm)/Argon gas(16 mm)/low-e(6 mm)]	1.29843
TGU [Clear(4 mm)/Argon gas(16 mm)/low-e(4 mm)/Argon gas(16 mm)/low-e(4 mm)]	0.70308

Source: Rezaei et al. (2017).

WINDOW RETROFITTING TECHNOLOGIES

Instead of replacing the single-pane windows with multi-pane windows, many advanced window retrofitting technologies have been developed to employ the single-pane window with low thermal transmittance and desirable optical properties. Currently, one widely used approach is storm windows. More attractively, coating thermal insulating and highly transparent materials onto glass panes is an effective and easy-handling approach to significantly decrease the U-factor of windows. We will introduce storm windows and few candidate coatings for single-pane window retrofitting, such as aerogel, low-e coating, polymer-air multilayers (PAM), in subsequent text.

Storm Windows

Storm window is an add-up layer mounted on the outmost surface of the window glass, which is much cheaper and easier to install than directly replacing or upgrading the entire windows. The windowpane can be made of inexpensive plastic sheets for short-term use, or high-end triple-track low-e coated glasses for long-term use. As for the frames, they can be made of cheap wood, aluminum (Al), plastics, and so on. By obtaining multiple-pane glazing, the total heat transfer coefficient can be mitigated. Nevertheless, the glass or special plastic sheets and frame for storm windows should be customized into size, which increases the retrofitting cost. In most cases, the storm windows cannot be opened, which is the biggest hindrance for storm windows to be widely used. Plus, water condensation may occur in storm windows, which is visually obstructive, presenting a poor visual impact. More maintenance may be needed if there is sealing problem in framing.

Aerogel Insulating Layer

Aerogel is an important candidate that is under extensive investigation as an attractive material to significantly increase the insulating property of windowpanes. It is manufactured mostly through sol-gel process, and the porous solid network with more than 90% empty volume typically is produced by extracting the liquid component of a gel through supercritical drying. It was first investigated by Kistler (1931) in 1931. The most developed and widely used aerogel for building applications is silica aerogel, formed from a silicon-based gel. Aerogel can reach an ultra-low thermal conductivity of around 0.012 W/m·K, VT more than 90%, and reflective index (RI) as low as 1.05 (Baetens et al., 2011). The extremely low thermal conductivity of aerogel is a result of Knudsen effect. When the cavity size is comparable to the mean free path of the gas, the cavity will hinder the movement of the gas

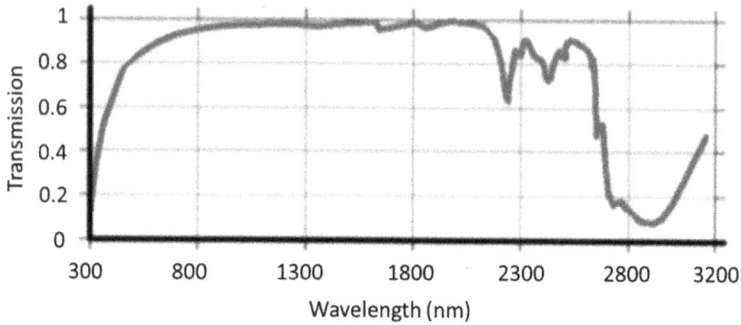

Figure 2.4 Light Transmittance of a Monolithic Aerogel Sample. *Source:* E-produced with permission from Baetens et al. (2011). © 2011 Elsevier.

particles, which results in decrease in the thermal conductivity. For instance, thermal conductivity of air is about 26 mW/m·K, while it can be reduced to about 5 mW/m·K in a pore of 30 nm diameter (Berge and Johansson 2012). Figure 2.4 shows the light transmittance spectrum of a typical monolithic aerogel sample (Baetens et al., 2011), which illustrates the high VT in the range 380–750 nm.

Thermal Performance of Aerogel

The average pore size of aerogel is in the range of 20–40 nm (Dorcheh and Abbasi, 2008). The nanopores in aerogels largely nullify thermal conduction as it is almost entirely composed of poorly conductive air; it inhibits convection as the pore sizes are less than the mean free path of air molecules. Knudsen effect is usually used to estimate the thermal conductivity of aerogels as (Kaganer, 1969; Bouquerel et al., 2012):

$$\lambda_g = \frac{\lambda_{g,0}}{1 + 2\xi K_n} \tag{2.5}$$

where, $\lambda_{g,0}$ is the conductivity of the gas without the aerogel, and ξ is coefficient characterizing the molecule-wall collision energy transfer (in)efficiency. $\xi = 3$ from Ref. (Heinemann et al., 1996). $\xi = 1.6$ from Ref. (Fricke et al., 2006). $\xi = 1.5$ from Ref. (Coquard and Quenard, 2007). K_n is Knudsen number, the ratio between the mean free path of air molecules and the diameter of the pores, s given as:

$$K_n = \frac{k_B T}{\sqrt{2}\pi d_g^2 P_g \delta} \tag{2.6}$$

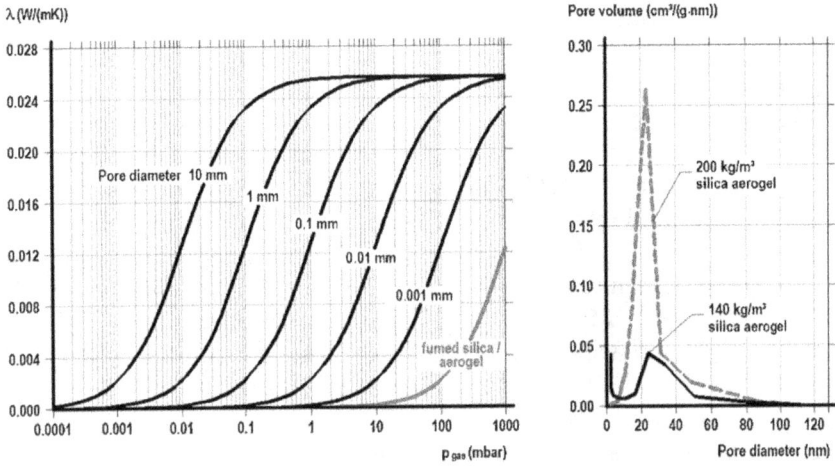

λ (W/(mK))   Pore volume (cm³/(g·nm))

Pore diameter 10 mm
1 mm
0.1 mm
0.01 mm
0.001 mm
fumed silica / aerogel
P_{gas} (mbar)

200 kg/m³ silica aerogel
140 kg/m³ silica aerogel
Pore diameter (nm)

Figure 2.5   The Thermal Conductivity of Air Changes with Varying of Air Pressure and the Average Pore Diameter. *Source:* E-produced with permission from Baetens et al. (2011). © 2011 Elsevier.

where, k_B is the Boltzmann constant, T is the temperature, d_g is the diameter of air molecules, P_g is the gas pressure, and δ is the pore size. Because of the nanometer range pore size δ (smaller than the mean free path of air to restrict convection) and the high porosity, thermal conductivity of aerogel can be lower than air ($\lambda_g < \lambda_{g,0}$) (Baetens et al., 2011). Well-developed silica aerogels are generally highly transparent exhibiting no color and no chromatic aberration, thanks to the nanopore size much smaller than the visual light wavelength [32,33]. Silica aerogel can provide single-pane widow with both thermal insulation and sound insulation without hurting its visual appearance. Figure 2.5 illustrates the thermal conductivity of silica aerogels with different pore diameter (Baetens et al., 2011). Table 2.4 shows some most widely used organic aerogels with their thermal conductivities and pore sizes.

Aerogel Fabrication

The most commonly used aerogel is made by silica, and it is synthesized typically through sol-gel process consisting of three major steps: gel preparation,

Table 2.4   Thermal Conductivities of the Organic Aerogels

Aerogels	Thermal Conductivity κ (W/m·K)	Pore Size (nm)
Polyurea	0.013 (Lee et al., 2009)	6.6–54
Resorcinol formaldehide	0.012 (Lu et al., 1992)	10–20
Polyciclopent adiene	0.017 (Lee and Gould, 2007)	–
Polyurethane	0.022 (Rigacci et al., 2004)	8.3–66.5
Cellulose	0.030 (Fischer et al., 2006)	1–100

gel aging, and critical drying or freeze-drying to avoid gel structure from collapse. At the beginning, the precursor materials are dispersed into the liquid (such as colloidal dispersion) and a continuous solid network will form in the liquid (Gurav et al., 2010). For some materials, cross-linked agencies are needed to improve the interaction between solid particles for the form of the gel (Aegerter et al., 2011; Capadona et al., 2006). The gelation duration is highly influenced by (i) chemical composition of the precursor and liquid, (ii) precursor and additives concentration, (iii) processing temperature, and (iv) pH value (Hdach et al., 1990; Hench and West, 1990; Mulik et al., 2007; Zhang et al., 2012). Many of them may still require additional curing after gelation, such as network perfection, to make the solid network strong enough (Hæreid et al., 1995; Omranpour and Motahari, 2013; Cheng and Iacobucci, 1988; Einarsrud et al., 2001; Dorcheh and Abbasi, 2008). When the gelation process is finished, to minimize the surface tension effects during the drying, supercritical carbon dioxide drying or freeze-drying can be applied, as shown in figure 2.6 (Barrios et al., 2019).

Aerogel for Window Insulation

At present, the most widely used aerogel for window application has overall density of 70–160 kg/m^3 (Baetens et al., 2011). However, because of its super high porosity, it exhibits extremely low fracture toughness, that is, the

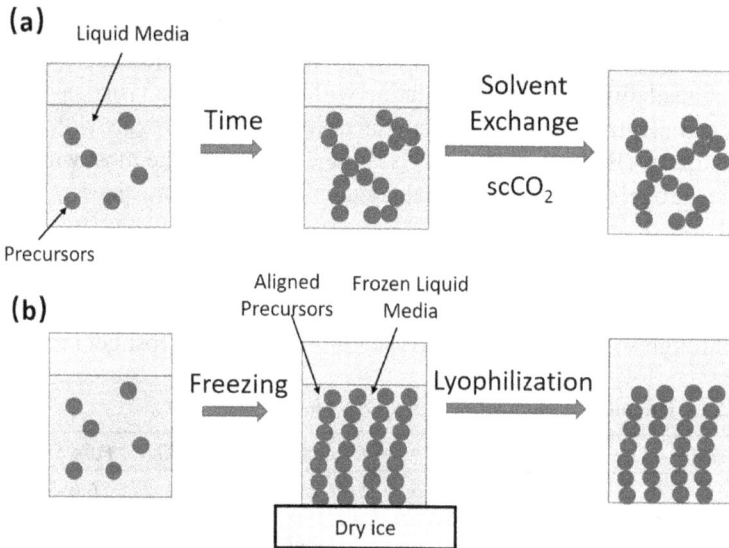

Figure 2.6 **The Schematic of (a) Supercritical Drying and (b) Freezing Drying Strategies.**
Source: Created by the authors.

capability to resist propagation of flaws in the material. Its fragility, with low Young's modulus and characteristic brittleness makes it far from being adopted in window insulation application [34]. To improve its structural integrity, the porosity should be reduced which will result in increase in its thermal conductivity. Once the porosity reaches ~50%, the thermal conductivity would be 0.04~0.1 W/m·K, much higher than air (Collins, 2019). Moreover, haze of most of silica aerogel is very large, due to the pore size distribution (Q. Liu et al. 2018; Maleki et al. 2014; Patel et al. 2009; Rubin and Lampert 1983), which is not desirable for window retrofitting, either. Any strengthening treatment or additives may hurt the VT or thermal insulation properties. A recent work (Liu et al., 2018) developed hybrid composite aerogels (TOCN-PMSQ) to address the problem of fragility by combing modified biopolymeric nanofibers with an inorganic polydioxanone network. By obtaining the optically isotropic or anisotropic hydrogels, organogels, and aerogels, the material can be highly transparent and has a relatively low haze. The effective thermal conductivity is lower than 0.025 W/m·K comparable to air, as illustrated in figure 2.7.

Low-E Coatings and Films

Concept and Application of Low-E Films

Window glass can emit the radiative energy in the form of long-wave, far-infrared energy depending on its surface temperature. Radiative heat losses through the window glass account for 60% of the total heat loss associated with windows (Rissman and Kennan, 2013). The emissivity of a material represents the ratio of heat which can be emitted out compared to that from

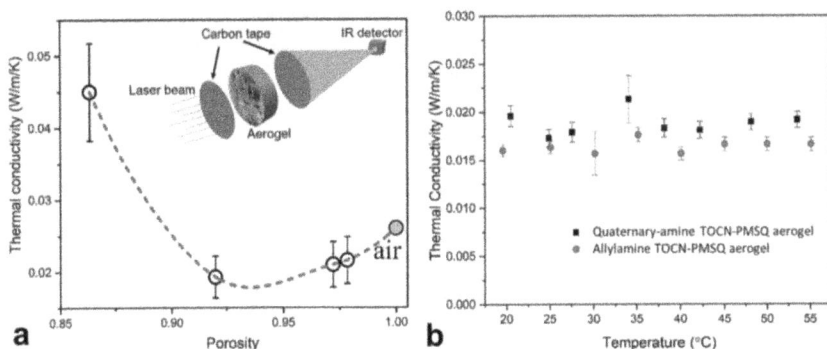

Figure 2.7   Thermal Conductivity of TOCN-PMSQ Aerogel as a Function of its Porosity and Temperature. *Source:* Reproduced with permission from Liu et al. (2018). © 2018 Elsevier.

Table 2.5 **Surface Emissivities of Common Materials**

Material Surface	Thermal Emissivity	Material Surface	Thermal Emissivity
Aluminum foil	0.03	Asphalt	0.88
Brick	0.9	Concrete, rough	0.91
Glass	0.84	Limestone	0.92
Silver	0.02		

Source: Handbook-Fundamentals and Edition (2009).

a blackbody, and ranges from 0 to 100%. The emissivity of typical common materials is shown in table 2.5 (Handbook-Fundamentals and Edition, 2009). Standard window glass pane's surface emissivity is ~0.84, indicating that 84% of the radiative thermal energy can be absorbed and emitted from the glass (Karlsson and Roos, 2001). Higher conductivity can lead to lower thermal emissivity, such as aluminum and silver in table 2.5. And multiple layers of aluminum foils are used in space technologies to provide good thermal insulating property. However, even though highly effective, this technology cannot be used for windowpanes as they are not transparent. Increasing thermal insulating property for windows is specially challenging as the ways to realize it often leads to reduced VT.

Low-e coating, a technology produced in the 1980s, was designed to prevent the long-wave radiative energy from passing through the window glass while allowing the passage of the visual light (DOE, 2014), as shown in figure 2.8. The VT can be higher than 70% while the infrared light transmittance is lower than 10%, which can block over 90% radiative heat transport.

In 1920s, Hagen and Rubens discovered the relationship between the emissivity, ε, and the material conductivity, σ, with the concentration of free elections (Glaser, 2000), indicating the higher the conductivity, the lower the emissivity. People also estimated the surface emissivity based on the material resistivity by following $\varepsilon = 0.0106R_a$, where R_a is the material resistivity (Glaser, 2000). Currently, semiconductor coating, for example, ITO (indium tin oxide) and FTO (fluorine-doped tine oxide) and the metallic coating are most adopted material for low-e coating. Their specifications are shown in table 2.6.

Hard coating, soft coating, and self-applicable films are three main methods to equip the window glass with low-e effect (Jelle et al., 2015; Schaefer et al., 1997; Rissman and Kennan, 2013; Martın-Palma et al., 1998). As for the hard coating, metal oxides, such as fluorine-doped tin oxide, can be applied during the production of the glass by chemical vapor deposition (CVD). Since the hard coating becomes a part of the glass, it can bear chemical erosion and mechanical wearing. While for soft coating, metal-based multilayer coatings can be applied to the glass after it is hardened. High-rate magnetron sputtering can be employed to coat the glass with multilayers of conductive

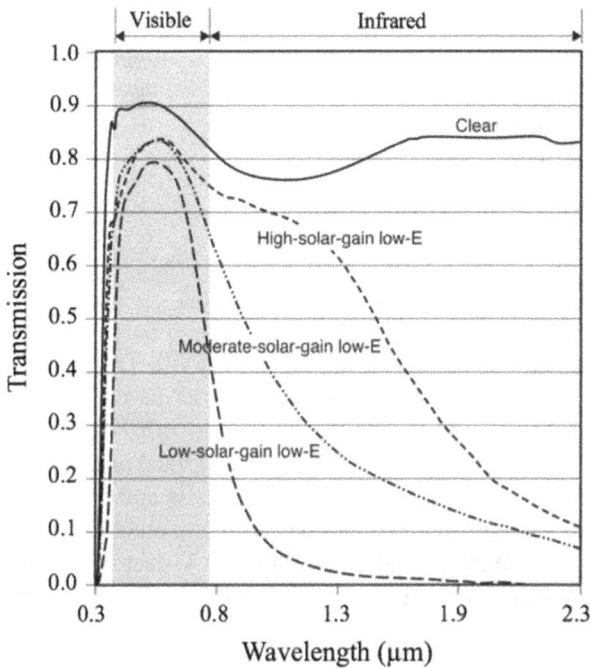

Figure 2.8   Spectral Transmittance Curves for Windows with Low-e Coatings. *Source*:
Reproduced with permission from Rezaei et al. (2017). © 2017 Elsevier.

metal and dielectric materials. The metal layers, generally with a thickness
<10 nm, has the capability to reflect the radiative thermal energy while the
dielectric layers, with a thickness <40 nm, serve as the anti-reflective coatings for improving the VT. Since the soft coating is not as robust as the hard
coatings, it is always applied onto the inside surfaces of double-pane or even
multi-pane windows. Table 2.7 shows VT and surface emissivities of some
common hard/soft coated glass (Grenadyorov et al., 2012; Solovyev et al.,

Table 2.6   The Common Low-e Materials for Window Coating

Specifications	ITO	FTO	Gold Layer Systems	Silver Layer Systems
Layer thickness (nm)	>20	>20	>6	>6
Sheet resistance R (Ω)	>8	>8	>5	>1
Visual transmittance (%)	>80	>80	>25	>85
Abrasion resistance	Desirable	Desirable	Good	Good
Chemical resistance	Good	Desirable	Moderate	Moderate
Total thickness (mm)	>0.3	>2	>0.3	>0.3

Source: Jelle et al. (2015).

Table 2.7   Multilayer Low-e Coatings and their Visible Transmittance and Emissivities

Low-e Multilayers	Visible Transmittance (%)	Emissivity (%)
Glass-ZnO-Ag-TiAO$_x$-SnO$_2$	86	10
Glass-SnO$_2$-Ag-NiCrO$_x$-SnO$_2$	84	15
Glass-SnO$_2$-TiAOx-Ag-TiAOx-SnO$_2$	84	12
Glass-SnBO$_2$-ZnO-Ag-NiCrOx-SnBO$_2$	85	9
Glass-Tungsten Oxide-Ag-Silicon buffer- Tungsten Oxide	86	10

Source: Grenadyorov et al. (2012), Solovyev et al. (2015), Demiryont et al. (2000), Schaefer et al. (1997).

2015; Demiryont et al., 2000; Schaefer et al., 1997). The most user- and eco-friendly low-e films are self-applicable polymer-based low-e films, which consist of multilayers of metallized polymers with back side coated with adhesives. It can be directly attached onto the surface of a window. However, since the self-applicable films are not mechanically and chemically durable or stable as hard coatings, these cannot be applied to outside surface of window glass to bear the weather conditions. Plus, the U-factor offered by attached polymer low-e films is not as low as that can be offered by the high-quality low-e glasses.

Challenges of Low-E Films

Despite above advantages, there are a couple of drawbacks of low-e coatings. At first, it won't block conductive heat loss through windows, which plays significant role in energy leakage through windows. Second, most low-e films have a natural grey haze and create a slight tint on windows, making the glass appear very different in color to clear glass and hurt the visibility (Schaefer et al. 1997; Szczyrbowski et al. 1989; Szanyi 2002). Because of its low VT, house owners will suffer from inadequate lighting inside. Moreover, once water condensation occurs on the low-e coating in cold weather, the low-e effectiveness is significantly reduced, making the low-e films noneffective ("Single-Pane Highly Insulating Efficient Lucid Designs (SHIELD) Program Overview" 2014). Thus, researchers have developed a new installation method to further improve the conductive heat resistance and minimize water condensation by elevating the low-e film from the window glass (Kou et al., 2019). The elevated low-e assembly is shorted as ELEA. Figure 2.9 shows a typical configuration of ELEA. Since the thermal conductivity of air is as low as 0.026 W/m·K, the overall thermal conductivity of the elevated low-e film together with the windowpane can be as low as 0.028 W/m·K. And the U-factor of single-pane windows can be reduced from 5.8 W/m²·K to lower than 2.8 W/m²·K, with a high VT (>70%) and low haze (<2%) (Kou et al., 2019).

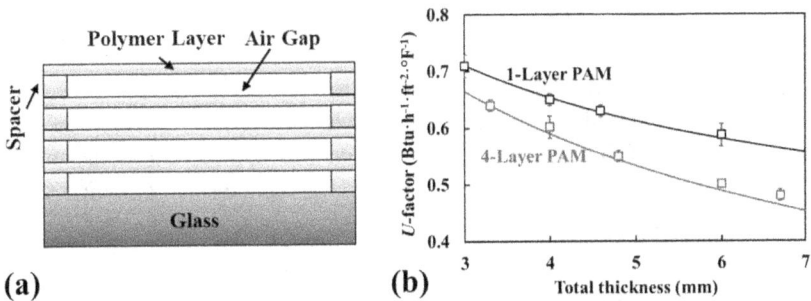

Figure 2.9 (a) A Schematic of the Polymer-Air Multilayer (PAM) and (b) the U-factor of PAM Coated Single-Pane Window as Function of the Total Thickness. 1 Btu/(h)=5.678 W/(m2·K). *Source*: Reproduced with permission from Kou et al. (2019). (c) 2019 Elsevier.

Polymer-Air Multilayers

Polymer-air multilayer structure was recently developed for single-pane window retrofitting. It consists of multiple layers of transparent polymer films separated by air gaps (Kou et al., 2020). Each polymer layer is about 0.01 to 0.25 mm thick and air gap thickness ranges from 0.1 to 1 mm. The total thickness of PAM varies from 3 to 10 mm. The layer separation was maintained by slightly stretching the polymer film or creating electrostatic forces between polymer films (Kou et al., 2019, 2020; Zhong et al., 2019). Since in each layer, the air gap thickness is less than 1 mm, the convective heat transfer can be significantly reduced compared to radiative and conductive heat transfer (Wright, 1996a). Multilayer of polymer film can significantly reduce radiative heat loss as each layer can block a portion of heat radiation. The gas between polymer films can be air with very low thermal conductivity ~0.026 W/m·K, and the overall thermal conductivity of PAM is as low as 0.03 W/m·K. The U-factor of single-pane window can be reduced from 5.8 W/m²·K to lower than 2.8 W/m²·K. While the VT is still higher than 70% and haze of the window is lower than 2%, as illustrated in figure 2.10.

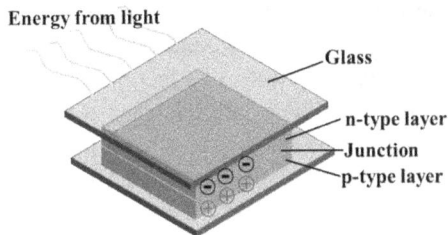

Figure 2.10 A Schematic of Photovoltaic Cell. *Source*: Created by the authors.

OTHER FUNCTIONAL COATINGS

Besides thermal insulating coating, there are also other coatings which can increase the sustainability of old window systems. In the subsequent sections, some developed functional materials for glazing application, such as anti-reflective coatings, self-cleaning coatings, photovoltaic (PV) glazing, will be discussed.

Self-Cleaning Coatings

Self-cleaning coatings can save energy by reducing energy consumption for cleaning. They can be divided into two main categories: hydrophobic and hydrophilic coatings. Both can clean themselves by the action of water. One is the by rolling water droplets and the latter one is based on sheeting water to carry away the contaminations. By applying self-cleaning coatings, the window maintenance will be minimized. VT and heat efficiency can be improved by reducing glass surface contamination (Giolando, 2016; Parkin and Palgrave, 2005).

People have developed many advanced techniques to produce hydrophobic and superhydrophobic surfaces. Hydrophobic surfaces can be produced by two main approaches: (1) roughening material surface, for example, hydrophobic surface can be manufactured by creating nanostructures and patterns (Jiang et al., 2014; Du et al., 2016; He et al., 2016); (2) modifying a rough surface with a material of low surface energy, such as coating TiO_2 along PDMS coating (Yang et al., 2016), using wet chemical reaction and hydrothermal reaction, (Pan et al., 2007), electrochemical deposition, (Li et al., 2007), etc. Hydrophilic coating is mainly based on coating TiO_2, ZnO, and WO_3 materials (Garlisi et al., 2015; Mellott et al., 2006; Liu et al., 2015).

Photovoltaic Glazing

Windows can not only be used for energy saving but also for energy harvesting. Window-integrated photovoltaics (WIPVs) have been well-developed to enable the active use of solar radiation and convert it into electrical energy (Li et al., 2009). The glazed window glass incorporates transparent semiconductor-based PV cells, known as solar cells. The solar cells are sandwiched between two glass panes with a special resin filled in to secure the solar cells. Each solar cell has two electrical connections to in order to form a connect with other solar and construct a whole system which can generate electrical energy (Ng and Mithraratne, 2014; Liao and Xu, 2015).

CONCLUSION AND FUTURE DEVELOPMENT

In conclusion, to mitigate energy loss through window glass and make buildings more energy sustainable, people have developed many advanced technologies to retrofit windows. Storm window is an add-up layer mounted either outside or inside of the entire window to make it functionalized as double-pane or triple-pane windows. By creating one or two more thick air gaps, the conduction heat transfer through the window glasses can be very much reduced. As for aerogel, it has lower thermal conductivity than air and high VT. However, its fragility and high haze hinder its widespread used. Low-e coatings can block most of thermal radiative heat transfer and have been used in most Energy Star windows; however, low-e films cannot block the conductive heat transfer and its low-e effect may get reduced if water condensation happens. Researchers have developed new installation approach to elevate the low-e coating from window glass to improve the conductive thermal resistance and hinder water condensation. Another new structure denoted as thermally insulating PAM structure has been developed with the functions to block both radiative and conductive heat transfer through window glass. It can reduce the U-factor of single-pane window from $1\sim1.2$ Btu/$(\mathrm{h}\cdot\mathrm{ft}^2\cdot{}^\circ\mathrm{F})$ to lower than 0.5 Btu/$(\mathrm{h}\cdot\mathrm{ft}^2\cdot{}^\circ\mathrm{F})$ at low cost. The VT and haze values are both desirable for window retrofitting. Besides of passive thermal insulating coatings, active energy-efficient materials have been developed to improve window energy sustainability. Self-cleaning coating can reduce energy consumption of window maintenance. Photovoltaic glazing can be applied to convert solar energy to electrical energy. In the future, both passive thermal insulating and active energy-converting technologies need to be further developed to improve the energy sustainability of windows that are the major energy leakage path for buildings at lower cost and lead the infrastructures into a new sustainable era.

REFERENCES

Abonyi, Brendan I, Juming Tang, and Charles G Edwards. 1999. "Evaluation of Energy Efficiency and Quality Retention for the Refractance Window (Drying System)." *Research Report.* December 30, 1999.

Aegerter, Michel A, N Leventis, and M M Koebel. 2011. *Advances in Sol-Gel Derived Materials and Technologies.* Aerogels Handbook; Springer, New York, NY, USA.

Afram, Abdul, and Farrokh Janabi-Sharifi. 2014. "Theory and Applications of HVAC Control Systems: A Review of Model Predictive Control (MPC)." *Building and Environment* 72: 343–355.

Arasteh, Dariush, Elizabeth Finlayson, Joe Huang, Charlie Huizenga, Robin Mitchell, and Mike Rubin. 1998. *State-of-the-Art Software for Window Energy-Efficiency Rating and Labeling.* Lawrence Berkeley National Lab., CA, USA.

Baetens, Ruben, Bjørn Petter Jelle, and Arild Gustavsen. 2011. "Aerogel Insulation for Building Applications: A State-of-the-Art Review." *Energy and Buildings* 43(4): 761–769.

Barrios, Elizabeth, David Fox, Yuen Yee Li Sip, Ruginn Catarata, Jean E. Calderon, Nilab Azim, Sajia Afrin, Zeyang Zhang, and Lei Zhai. 2019. "Nanomaterials in Advanced, High-Performance Aerogel Composites: A Review." *Polymers* 11(4): 726.

Berge, Axel, and P Ä R Johansson. 2012. *Literature Review of High Performance Thermal Insulation*. Göteborg, Sweden: Chalmers University of Technology.

Bergman, Theodore L, Frank P Incropera, David P DeWitt, and Adrienne S Lavine. 2011. *Fundamentals of Heat and Mass Transfer*. Hoboken, NJ: John Wiley & Sons.

Bouquerel, Mathias, Thierry Duforestel, Dominique Baillis, and Gilles Rusaouen. 2012. "Heat Transfer Modeling in Vacuum Insulation Panels Containing Nanoporous Silicas—A Review." *Energy and Buildings* 54: 320–336.

Capadona, Lynn A, Mary Ann B Meador, Antonella Alunni, Eve F Fabrizio, Plousia Vassilaras, and Nicholas Leventis. 2006. "Flexible, Low-Density Polymer Crosslinked Silica Aerogels." *Polymer* 47(16): 5754–5761.

Carmody, John, Stephen Selkowitz, Eleanor Lee, Dariush Arasteh, and Todd Willmert. 2004. *Window System for High-Performance Buildings*. New York, NY: W. W. Norton & Company.

Cheng, Chung-Ping, and Paul A Iacobucci. 1988. "Inorganic Oxide Aerogels and Their Preparation." Google Patents.

Collins, Richard. 2019. "Why Isn't the Aerogel Industry Booming?" https://www.idt echex.com/fr/research-article/why-isnt-the-aerogel-industry-booming/16671.

Coquard, Rémi, and Daniel Quenard. 2007. "Modeling of Heat Transfert in Nanoporous Silica-Influence of Moisture." In *Proceedings of the 8th International Vacuum Insulation Symposium*, pp. 1–13. Wurzburg, September 18–19, 2007.

Demiryont, Huyla, Huseyin Parlar, Ayse Ersoy, and Ender Aktulga. 2000. "Anti-Solar and Low Emissivity Functioning Multi-Layer Coatings on Transparent Substrates." Google Patents.

DoE, U. S. 2011. *Buildings Energy Databook*. Energy Efficiency & Renewable Energy Department.

DOE, U. S. 2014. *Windows and Building Envelope Research and Development: Roadmap for Emerging Technologies*. US Department of Energy, Buildings Technologies Office, Washington, DC.

Dorcheh, A Soleimani, and M H Abbasi. 2008. "Silica Aerogel; Synthesis, Properties and Characterization." *Journal of Materials Processing Technology* 199(1–3): 10–26.

Du, Xin, Yi Xing, Xiaoyu Li, Hongwei Huang, Zhi Geng, Junhui He, Yongqiang Wen, and Xueji Zhang. 2016. "Broadband Antireflective Superhydrophobic Self-Cleaning Coatings Based on Novel Dendritic Porous Particles." *RSC Advances* 6(10): 7864–7871.

Einarsrud, M-A, Elin Nilsen, Arnaud Rigacci, Gérard Marcel Pajonk, S Buathier, D Valette, M Durant, B Chevalier, Peter Nitz, and Françoise Ehrburger-Dolle. 2001. "Strengthening of Silica Gels and Aerogels by Washing and Aging Processes." *Journal of Non-Crystalline Solids* 285(1–3): 1–7.

Environmental Protection Agency. 2015. "Energy Star Performance Criteria for Windows, Doors, and Skylights." https://www.energystar.gov/products/building_products/residential_windows_doors_and_skylights/key_product_criteria.

Fischer, Florent, Arnaud Rigacci, R Pirard, Sandrine Berthon-Fabry, and Patrick Achard. 2006. "Cellulose-Based Aerogels." *Polymer* 47(22): 7636–7645.

Fricke, Jochen, E Hümmer, H-J Morper, and P Scheuerpflug. 1989. "Thermal Properties of Silica Aerogels." *Le Journal de Physique Colloques* 50(C4): C4–C87.

Fricke, Jochen, Hubert Schwab, and Ulrich Heinemann. 2006. "Vacuum Insulation Panels–Exciting Thermal Properties and Most Challenging Applications." *International Journal of Thermophysics* 27(4): 1123–1139.

Garlisi, C, G Scandura, A Alabi, O Aderemi, and G Palmisano. 2015. "Self-Cleaning Coatings Activated by Solar and Visible Radiation." *Journal of Advanced Chemical Engineering* 5(1): 1–3.

Giolando, Dean M 2016. "Transparent Self-Cleaning Coating Applicable to Solar Energy Consisting of Nano-Crystals of Titanium Dioxide in Fluorine Doped Tin Dioxide." *Solar Energy* 124: 76–81.

Glaser, Hans Joachim. 2000. *Large Area Glass Coating*. Von Ardenne Anlagentechnik GmbH.

Grenadyorov, Alexandr S, Sergey V Rabotkin, and A S Parnyugin. 2012. "The Effectiveness of Low-Emissivity Coating on a Polymer Film." In *2012 7th International Forum on Strategic Technology (IFOST)*, pp. 1–4. IEEE. 18–21 September 2012, Tomsk, Russia

Griffith, B T, and D Arasteh. 1992. "Gas-Filled Panels: A Thermally Improved Building Insulation." Thermal Performance of the Exterior Envelopes of Buildings V Conference Proceedings, Clearwater Beach, FL, December, 1992.

Gurav, Jyoti L, In-Keun Jung, Hyung-Ho Park, Eul Son Kang, and Digambar Y. Nadargi. 2010. "Silica Aerogel: Synthesis and Applications." *Journal of Nanomaterials* Volume 2010: 1–11.

Hæreid, S, J Anderson, M A Einarsrud, D W Hua, and D M Smith. 1995. "Thermal and Temporal Aging of TMOS-Based Aerogel Precursors in Water." *Journal of Non-Crystalline Solids* 185(3): 221–226.

Handbook-Fundamentals, ASHRAE, and S I Edition. 2009. "Atlanta: American Society of Heating, Refrigerating and Air-Conditioning Engineers." *Inc.(See Page 14.14 for Summary Description of RP-1171 Work on Uncertainty in Design Temperatures)*.

Hdach, H, T Woignier, J Phalippou, and G W Scherer. 1990. "Effect of Aging and PH on the Modulus of Aerogels." *Journal of Non-Crystalline Solids* 121(1–3): 202–205.

He, Zhiwei, Zhiliang Zhang, and Jianying He. 2016. "CuO/Cu Based Superhydrophobic and Self-Cleaning Surfaces." *Scripta Materialia* 118: 60–64.

Heinemann, Ulrich, Roland Caps, and Jochen Fricke. 1996. "Radiation-Conduction Interaction: An Investigation on Silica Aerogels." *International Journal of Heat and Mass Transfer* 39(10): 2115–2130.

Hench, Larry L, and Jon K West. 1990. "The Sol-Gel Process." *Chemical Reviews* 90(1): 33–72.

Horowitz, Flavio, Marcelo B Pereira, and Giovani B de Azambuja. 2011. "Glass Window Coatings for Sunlight Heat Reflection and Co-Utilization." *Applied Optics* 50(9): C250–C252.

Howell, John R, M. Pinar Menguc, and Robert Siegel. 2015. *Thermal Radiation Heat Transfer*. CRC Press, Boca Raton, FL.

Incropera, Frank P, Adrienne S Lavine, Theodore L Bergman, and David P DeWitt. 2007. *Fundamentals of Heat and Mass Transfer*. Wiley, Hoboken, NJ.

Jelle, Bjørn Petter, Simen Edsjø Kalnæs, and Tao Gao. 2015. "Low-Emissivity Materials for Building Applications: A State-of-the-Art Review and Future Research Perspectives." *Energy and Buildings* 96: 329–356.

Jiang, Y D, M. Kitada, M White, T Fitz, and A T Hunt. 2014. "Development of Durable Nanostructured Superhydrophobic Self-Cleaning Surfaces on Glass Substrates." *Journal of Food Process Preserve* 38(3): 1321–1329.

Kaganer, Mikhail Grigorevich. 1969. "Thermal Insulation in Cryogenic Engineering." Jerusalem, Israel Program for Scientific Translations, 1969.

Karlsson, J, and A Roos. 2001. "Annual Energy Window Performance vs. Glazing Thermal Emittance—The Relevance of Very Low Emittance Values." *Thin Solid Films* 392(2): 345–348.

Kistler, Samuel Stephens. 1931. "Coherent Expanded Aerogels and Jellies." *Nature* 127(3211): 741.

Klems, J 2003. "Measured Summer Performance of Storm Windows." *Lawrence Berkeley National Laboratory Report. Q9.*

Kou, Rui, Ying Zhong, Jeongmin Kim, Qingyang Wang, Meng Wang, Renkun Chen, and Yu Qiao. 2019. "Elevating Low-Emissivity Film for Lower Thermal Transmittance." *Energy and Buildings* 193: 69–77.

Kou, Rui, Ying Zhong, Qingyang Wang, Jeongmin Kim, Renkun Chen, and Yu Qiao. 2020. "Thermal Insulating Polymer-Air Multilayer for Window Energy Efficiency." *ArXiv Preprint ArXiv:2005.14395.*

Kou, Rui, Ying Zhong, and Yu Qiao. 2019. "Effects of Anion Size on Flow Electrification of Polycarbonate and Polyethylene Terephthalate." *Applied Physics Letters* 115(7): 73704.

Kou, Rui, Ying Zhong, and Yu Qiao. 2020. "Flow Electrification of Corona-Charged Polyethylene Terephthalate Film." *ArXiv Preprint ArXiv:2005.14385.*

Lee, Je Kyun, and George L Gould. 2007. "Polydicyclopentadiene Based Aerogel: A New Insulation Material." *Journal of Sol-Gel Science and Technology* 44(1): 29–40.

Lee, Je Kyun, George L Gould, and Wendell Rhine. 2009. "Polyurea Based Aerogel for a High Performance Thermal Insulation Material." *Journal of Sol-Gel Science and Technology* 49(2): 209–220.

Li, Danny H W, Tony N T Lam, Wilco W H Chan, and Ada H L Mak. 2009. "Energy and Cost Analysis of Semi-Transparent Photovoltaic in Office Buildings." *Applied Energy* 86(5): 722–729.

Li, Ying, Wen-Zhi Jia, Yan-Yan Song, and Xing-Hua Xia. 2007. "Superhydrophobicity of 3D Porous Copper Films Prepared Using the Hydrogen Bubble Dynamic Template." *Chemistry of Materials* 19(23): 5758–5764.

Liao, Wei, and Shen Xu. 2015. "Energy Performance Comparison Among See-Through Amorphous-Silicon PV (Photovoltaic) Glazings and Traditional Glazings under Different Architectural Conditions in China." *Energy* 83: 267–275.

Liu, Hang, Yajie Chen, Guohui Tian, Zhiyu Ren, Chungui Tian, and Honggang Fu. 2015. "Visible-Light-Induced Self-Cleaning Property of $Bi_2Ti_2O_7$-TiO_2 Composite Nanowire Arrays." *Langmuir* 31(21): 5962–5969.

Liu, Qingkun, Allister W Frazier, Xinpeng Zhao, A Joshua, Andrew J Hess, Ronggui Yang, and Ivan I Smalyukh. 2018. "Flexible Transparent Aerogels as Window Retrofitting Films and Optical Elements with Tunable Birefringence." *Nano Energy* 48: 266–274.

Lu, Ning. 2012. "An Evaluation of the HVAC Load Potential for Providing Load Balancing Service." *IEEE Transactions on Smart Grid* 3(3): 1263–1270.

Lu, X, M C Arduini-Schuster, J Kuhn, O Nilsson, J Fricke, and R W Pekala. 1992. "Thermal Conductivity of Monolithic Organic Aerogels." *Science* 255(5047): 971–972.

Maleki, Hajar, Luisa Durães, and António Portugal. 2014. "An Overview on Silica Aerogels Synthesis and Different Mechanical Reinforcing Strategies." *Journal of Non-Crystalline Solids* 385: 55–74.

Martın-Palma, R J, L Vazquez, and J M Martınez-Duart. 1998. "Silver-Based Low-Emissivity Coatings for Architectural Windows: Optical and Structural Properties." *Solar Energy Materials and Solar Cells* 53(1–2): 55–66.

Mellott, N P, C Durucan, Carlo G Pantano, and M Guglielmi. 2006. "Commercial and Laboratory Prepared Titanium Dioxide Thin Films for Self-Cleaning Glasses: Photocatalytic Performance and Chemical Durability." *Thin Solid Films* 502(1–2): 112–120.

Mulik, Sudhir, Chariklia Sotiriou-Leventis, and Nicholas Leventis. 2007 "Time-Efficient Acid-Catalyzed Synthesis of Resorcinol–Formaldehyde Aerogels." *Chemistry of Materials* 19(25): 6138–6144.

Ng, Poh Khai, and Nalanie Mithraratne. 2014. "Lifetime Performance of Semi-Transparent Building-Integrated Photovoltaic (BIPV) Glazing Systems in the Tropics." *Renewable and Sustainable Energy Reviews* 31: 736–745.

Omranpour, Hosseinali, and Siamak Motahari. 2013. "Effects of Processing Conditions on Silica Aerogel during Aging: Role of Solvent, Time and Temperature." *Journal of Non-Crystalline Solids* 379: 7–11.

Pan, Qinmin, Haizu Jin, and Hongbo Wang. 2007. "Fabrication of Superhydrophobic Surfaces on Interconnected $Cu(OH)_2$ Nanowires via Solution-Immersion." *Nanotechnology* 18(35): 355605.

Parkin, Ivan P, and Robert G Palgrave. 2005. "Self-Cleaning Coatings." *Journal of Materials Chemistry* 15(17): 1689–1695.

Patel, Rakesh P, Nirav S Purohit, and Ajay M. Suthar. 2009. "An Overview of Silica Aerogels." *International Journal of ChemTech Research* 1(4): 1052–1057.

Rezaei, Soroosh Daqiqeh, Santiranjan Shannigrahi, and Seeram Ramakrishna. 2017. "A Review of Conventional, Advanced, and Smart Glazing Technologies and Materials for Improving Indoor Environment." *Solar Energy Materials and Solar Cells* 159: 26–51.

Rigacci, Arnaud, J C Marechal, Monique Repoux, Maryline Moreno, and Patrick Achard. 2004. "Preparation of Polyurethane-Based Aerogels and Xerogels for Thermal Superinsulation." *Journal of Non-Crystalline Solids* 350: 372–378.

Rissman, Jeffrey, and Hallie Kennan. 2013. "Low-Emissivity Windows." *American Energy Innovation Council March* 1: 12.

Rubin, Michael, and Carl M. Lampert. 1983. "Transparent Silica Aerogels for Window Insulation." *Solar Energy Materials* 7(4): 393–400.

Schaefer, C, G Bräuer, and J Szczyrbowski. 1997. "Low Emissivity Coatings on Architectural Glass." *Surface and Coatings Technology* 93(1): 37–45.

Simmler, Hans, S Brunner, U Heinemann, H Schwab, K Kumaran, Ph Mukhopadhyaya, D Quénard, H Sallée, K Noller, and E Küküpinar-Niarchos. 2005. "Vacuum Insulation Panels: Study on VIP-Components and Panels for Service Life Prediction of VIP in Building Applications (Subtask A)." *IEA/ECBCS Annex* 39: 1–153.

"Single-Pane Highly Insulating Efficient Lucid Designs (SHIELD) Program Overview." 2014. pp. 1–13.

Solovyev, A A, S V Rabotkin, and N F Kovsharov. 2015. "Polymer Films with Multilayer Low-E Coatings." *Materials Science in Semiconductor Processing* 38: 373–380.

Szanyi, János. 2002. "The Origin of Haze in CVD Tin Oxide Thin Films." *Applied Surface Science* 185(3–4): 161–171.

Szczyrbowski, J, A Dietrich, and K Hartig. 1989. "Bendable Silver-Based Low Emissivity Coating on Glass." *Solar Energy Materials* 19(1–2): 43–53.

US Department of Energy. 2014. *Research and Development Roadmap for Emerging HVAC Technologies*. US Department of Energy, Buildings Technologies Office, Washington, DC, no. October, p. 121.

Wright, John L 1996a. "A Correlation to Quaatify Transfer BetweeaV Ertical Convective Heat Window Glazings." In *ASHRAE Transactions, Pt 106*. American Society of Heating, Refrigerating and Air-Conditioning Engineers.

Wright, John L 1996b. "A Correlation to Quantify Convective Heat Transfer between Vertical Window Glazings." *ASHRAE Transactions* 102, Part 1: 940–946.

Yang, Shu-Jing, Xin Chen, Bing Yu, Hai-Lin Cong, Qiao-Hong Peng, and Ming-Ming Jiao. 2016. "Self-Cleaning Superhydrophobic Coatings Based on PDMS and TiO_2/SiO_2 Nanoparticles." *Integrated Ferroelectrics* 169(1): 29–34.

Zhang, Jing, Yewen Cao, Jiachun Feng, and Peiyi Wu. 2012. "Graphene-Oxide-Sheet-Induced Gelation of Cellulose and Promoted Mechanical Properties of Composite Aerogels." *The Journal of Physical Chemistry C* 116(14): 8063–8068.

Zhong, Ying, Rui Kou, Meng Wang, and Yu Qiao. 2019. "Electrification Mechanism of Corona Charged Organic Electrets." *Journal of Physics D: Applied Physics* 52(44): 445303.

Chapter 3

A Critical Review of Urban Energy Solutions and Practices

Negin Minaei

ABBREVIATIONS

AC: Air conditioner
EIA: Environmental Impact Assessment
GHG: Greenhouse emissions
HVAC: Heating, ventilation, and air conditioning
LEED: Leadership in Energy and Environmental Design
SDGs: Sustainable Development Goals
SRFs: Solid recovered fuels
UHI: Urban heat islands
UN: United Nations
UNDP: United Nations Development Programme
UNEP: UN Environment Programme
VOCs: Volatile organic compounds
GEMET: General Multilingual Environmental Thesaurus

INTRODUCTION

In this chapter, I revisit the definition of Clean Tech and investigate some common practices and products in engineering field that have newly emerged but are not compliant with the real concept of sustainability and Sustainable Development Goals.[1] Although the objectives of these solutions and practices are to reduce greenhouse gas (GHG) emissions, decrease carbon footprint, and ease sustainable living for public, these unsustainable engineering solutions impose environmental burdens[2] and put Earth's resources under pressure. I briefly mention different examples of these unsustainable practices and

53

environmental burdens, for example, urban heat islands (UHIs)[3] and explain why it is tightly woven to a so-called green solution in construction industry. I also explain why a Systems Thinking approach is necessary in dealing with cities and our environment to help us see the bigger picture and prevent us from inventing tunnel-vision solutions. A careful observation and a thorough critical study can identify some issues that do not work in an urban context. Most of these solutions are reinventions of original good old ideas used by our ancestors for a long time without any negative impacts on our planet.

This chapter does not aim to review literature on power generation or engineering design methods. It aims to attract the attention of engineers and designers as well as cities and municipalities who are Clean Tech users to see the broader impacts of each design and technology on a larger scale. It aims to shed light on some of the so-called environmental-friendly solutions currently promoted by different industries and help cities move toward sustainability. These examples range from power generation to waste management and construction industry. They were chosen from across the world because they were promoted globally and since cities are competing to become smart cities instead of sustainable smart cities[4] (Minaei, 2017, 2020a), they look for the state-of-the-art solutions.

METHODOLOGY: SYSTEM THINKING
AND DESIGN THINKING

For two years, I taught System Thinking and Design Thinking to engineering students of different programs and asked them to design products that could solve urban challenges. I have seen that often students and industry emphasize on a dimension that support their ideas but that does not necessarily mean that they have looked at the concept and system comprehensively. This can be because of a limited scope and a tunnel vision. I use two methods that provide us with the necessary tools to efficiently design engineering solutions. Coughlan and Ponto also recommended using System Thinking and Design Thinking conjointly in designing for sustainability transition while they comprehensively explained how these two differ and how they complete each other (Gaziulusoy, 2012). Here, I briefly explain them to illustrate the necessity of applying them in studying cities as complex systems, for example, power generation in cities, waste management in cities, and role of air conditioners (AC) in green construction industry and on the urban environments.

System Thinking as a discipline provides tools and skills to help us manage complex systems and uncertain situations that are often difficult to understand or predict and therefore it is not easy to find answers to their problems. In a complex system, there are large numbers of components so the behavior of the system is not predictable as it is multilayered and many interactions among different factors are happening simultaneously. The famous example is ecosystem in which everything works in a particular way. Cities are often more complex as we have humans who are the most unpredictable factors in a system. System Thinking tools help us to see the bigger picture with all factors and their interactions. It can prevent short-sightedness and allow us to predict for future.

Design Thinking has six phases: empathize, define, ideate, prototype, test, and implement. It starts from empathizing with your user and identifying what their needs are. Define a problem clearly so you can ideate many solutions and generate ideas. Then decide the ways you want to solve the problem and choose your design. The next step is to prototype your chosen design and test it. Return to your users and get feedback from them. Now is the time to put your vision into effect and implement it. Some Design Thinking users apply an alternative model which has one extra phase titled storey telling and it comes before implementation. It is believed that the designers should know why their work matters. Every design should have a purpose and should solve a problem; it is the most essential rationale for inventing a product. We shall distance ourselves from luxury and gadget products and designs that do not serve any real purpose.

CLEAN TECHNOLOGY

Although many innovators, engineers, and industry partners think they are at the forefront of clean technologies, some of the products or services are in fact unsustainable and should not be considered as clean or green. I recognize the Eco Canada's (2020) definition for Clean Tech as "any type of process, product or service that improves performance with lower costs while negative environmental impacts are minimized, and natural resources are responsibly and efficiently used." Many Clean Tech users are providers of urban services such as natural resources, utilities including alternate energies, construction of mainly green buildings, sustainable transport, manufacturing, waste reduction, lifecycle management and pollution control, and research and development. This clearly illustrates the reason for United Nations (UN) to rightly assign a goal to it, the Goal #11, Sustainable Cities and Communities as cities' contribution to sustainable development can be significant and municipalities are often the number one user of most of these green technologies. In Canada alone, 38% of the Clean Tech applications belong to pipeline transportation, 36% to utilities, 22% to rail and water transportation, and 20% to oil and gas extraction (Eco Canada, 2020).

SDG #11 AND URBAN HEAT ISLANDS

Cities are responsible for more than 70% of the world's CO_2 emissions (UNHABITAT, 2016a). Urban population growth and increased demand for energy in cities have led to human enhanced greenhouse effects, climate change, and consequently UHIs. In 2014, 87% of disasters were climate-related (UNHABITAT, 2016a). That shows the importance of taking right actions by governments. SDG Goal #11 attempts to assist national and local governments to monitor and report the indicators for the sustainable cities and community framework (UNHABITAT, 2016b). Looking for solutions to decrease the carbon footprint and conserve energy, research institutes and industry have collaborated and proposed variety of solutions from which some have been opted by cities. There are myths that have found their way into energy policies and even bylaws and have been amended by cities, but their efficiency is questionable. For example, application of white roofs in all areas regardless of their climate is not the right answer particularly in cold climates which may have the reverse effect. LEED, the U.S. Department of Energy, the Cool Roof Rating Council and ENERGY STAR program believe that white reflective roofing systems are more efficient in reducing global warming and UHIs, as well as are more cost-effective than dark roofing even in northern cities; that cannot be the case in winter for areas with cold climate (Ibrahim, 2013; Luna, 2015) such as Canada. These types of solutions should be conducted based on the location data including the climate. The importance of urban climatology for tropical areas has been highlighted in publications of Gonzales et al. (2005) and Murphy et al. (2011). While white roofs can be good choices in hot and arid climate to keep the urban environments cool, they are not logical choices for cities with cold climates such as Toronto. Cool roofs can keep buildings a little cooler while the overall number of hot days of summer in Toronto does not surpass even 30 days. Accurate calculations and studies are needed to determine for which temperature range it could optimally function.

On the contrary, ecological benefits of green roofs include their contribution to UHIs by decreasing the over roof temperature considerably; thus, green roofs create ambient temperatures because of the microclimate they create. They play many roles including: (1) storm water management, (2) pressure reduction from city's drainage system, (3) air pollution reduction by filtering some toxic particles, (4) improving air quality, (5) decreasing noise pollution and helping soundproofing, and (6) providing a green space over a roof top which can not only be used for urban food production but also as a refuge space to calm down and relax.

One of the most known impacts of climate change on cities and urban environments is the phenomena of UHIs. One of the diagnostics of climate change is its impacts on environment and particularly ecosystems. Changes in vegetation phenology can be considered as one. UHIs can potentially change the growing season and the vegetation phenology (Melaas et al., 2016) and that includes the cycle of emergence, development, and senescence of leaves and it is controlled by temperature (Chemielewski and Rotzer, 2001). One solution to prevent UHI is to use green facades to lower buildings' temperature in cities. Increasing vegetation is the most efficient solution identified to decrease the heat islands in cities and urban areas. A study recommended installing solar thermal panels on facades to benefit from the solar potentials of facades (Mohajeri et al., 2016, p. 481). In this study, two important contributing factors meaning trees and material of pavements and surfaces were omitted. Despite the positive benefits of the solar panels, adding dark solar panels to facades absorbs more heat and has an adverse effect unless clear or light color solar panels are produced. In addition, even solar panels need to stay cool to produce power efficiently and that is the reason hybrid roofs covered with vegetation are often recommended rather than white roofs to keep photovoltaic panels at their optimum performance.

Air-Tight Buildings and Air Conditioners

A review of studies on UHI shows that the most three important factors that contribute to the UHI are: *landscaping, using albedo materials on external surfaces of buildings and urban areas, and using natural ventilation* (Shahmohamadi et al., 2011). Using ACs as a major factor in buildings backfires. The best solution is to benefit from natural ventilation. Unfortunately, even green, zero carbon, and passive house buildings suggest using them without seeing the bigger picture. I have comprehensively discussed all styles of sustainable architecture in my former chapter under "Sustainable Architectural Design and Construction" (Minaei, 2020a). A single building equipped with heating, ventilation, and air conditioning (HVAC) or AC can perform well in terms of energy efficiency; but we should be mindful of the impacts that it has on the city. That is the reason that we should use System Thinking approach in our studies to see a building in both architectural scale (a single building) and urban scale (in relation to its urban context and urban ecology).

Closing all ventilation gaps and air-tightening buildings and only relying on a 24/7 HVAC and AC is not a sustainable and resilient option since it consumes energy all the time and in case of power outage, there is no reliable source for ventilation, heating, and cooling. Studies have shown that ACs cool down the indoor air, increase the indoor air quality but pollute and warm up the outdoor air which decreases the outdoor air quality and eventually leads to UHIs. Lower air quality contributes to more UHIs effects and makes it harder for people to breathe in the surrounding areas.

Also having access to clean and fresh air is important. One lesson learned from the Spanish Flu pandemic was to benefit from natural air circulation and solar irradiation through windows as they can help cleansing an environment. Historical evidence of former pandemics has proved fresh air and sunshine can speed up recovery from viral infections (Hobday, 2020). North American adults spend about 87% of their time indoors, and operable windows or proper ventilation systems are crucial to ensure they decrease the amount of biological and chemical contaminants (Dales et al., 2008) that occupants are exposed to. Windows' first important function is to provide natural air circulation and ventilation which often decreases the amount of CO_2, volatile organic compounds, and formaldehyde and increases the indoor air quality (The Well Standard, 2018).

Windows and Sunshine as a Natural Thermal Heating Source

Every home needs operable windows, sunshine, light, and view. South-facing windows in cold climates can provide natural free heating with a thermal comfort while saving energy which is the ultimate goal for sustainable construction and green buildings.[5] Natural air and sunshine are necessary for health and well-being of both children and adults.

In his book *Happy City*, Charles Montgomery mentioned the local obsession with views and wrote: "almost nobody in the city wants to face south, where the sun occasionally appears through the rain clouds" (Montgomery, 2013, p. 118, 2). It is surprising that Vancouverites prefer views to north and west rather than having windows to the south (Minaei, 2020b)! This is another example of the wrong practices. Architecturally speaking, not opting for south-facing windows in cold climate to get sunshine is simply wrong, far from energy-efficient, and unhealthy. If views of mountains in the north side are beautiful, windows on north side could be kept to provide nice view and natural light and some windows on the south side could be designed to provide sunshine and heat. Unless a building is in a hot climate and needs to avoid direct sunshine from south, in most climates, architects benefit from the south-side windows to warm up buildings and at the same time bring natural light inside throughout a day which can save electricity. In passive solar buildings, designers try to benefit from sunshine as much as possible employing different methods such as "direct gain," "indirect gain," and "isolated gain" (Minaei, 2020a). In passive house buildings, we try to minimize the size and number of openings in a building in order to minimize the energy waste.

Environmental psychology research proved having nice views, particularly natural views, can have restorative effects and increase concentration. In times of pandemic lock downs, we appreciate the value of windows in our buildings. Operable windows are necessary for our health and well-being, not only because they provide natural ventilation, but also because they give us the feeling of being in control; and we know that a good quality of life is related to the feeling of being in control which directly impacts our mental health (Connell et al., 2012).

Eco-Walls, Green Walls, or Living Walls

Eco-walls, green walls, and living walls are different terms that are used interchangeably. In green urbanism and biophilic cities, which are about connecting nature and people, all these terms are often used. For example, a living wall has the plants rooted on the wall. A green wall can be any wall that is covered with green plants—either they grow in the ground and go up on the wall or they root on the wall. A green wall can be placed outdoor, for instance, a façade that is covered with plants can be called a green wall. Eco-walls are also covered with plants, but they have an extra function which is to reduce noise, they function as a baffle. They have lots of planters and are installed between a noisy area like railway or highway and a residential neighborhood to break the noise. In this section, I discuss the living walls that are installed by a company in an indoor environment and plants grow their roots on the wall.

While the living walls are considered a sustainable source of oxygen by purifying the indoor air, most of the existing commercial products are not sustainable. The products often have:

- Water pumps that work 24/7 and consume electricity;
- In most cases, they are connected to a tap water and waste clean drinking water;
- They have artificial lights on 24/7 that consume electricity;
- Some of their plants do not belong to the air-purifying category and only are used for beautification, often these plants are tropical and imported from other countries which add to their carbon footprint due to long-distance shipping.
- They need regular maintenance visits and often they get algae in their reservoirs which is not healthy to breath. That means extra costs are forced on buyers.

The more sustainable option is to place these eco-walls or simply some air-purifying plants where they are exposed to natural light, employ mechanical systems to water them with harvested rainwater instead of wasting clean drinking water. Only 3% of the water in the world is drinkable (CCAO,

2019) and we should not waste it as much as possible. Our civil engineering students tried to solve these problems in their CAPSTONE final project in 2017. For example, by collecting rainwater and benefiting from sunshine, it is possible to omit the energy consumption and save clean drinking water.

Wood as a Construction Material in Green Buildings

Recent trends and policies in green buildings suggest measuring the carbon footprint of buildings. GHG emissions are not easy to measure for all industries so carbon footprint has found popularity among some industries like construction and design. A recent approach points to wood as a sustainable construction material. In this section, I debate this view by bringing some rationales.

Since soil, sand, and metal resources are limited and are finishing, we cannot continue building structures made of concrete and steel with the same pace as before. Instead of investing on recycling construction material, the construction industry is opting for wood as a sustainable source. Their rational is that since trees capture the CO_2, the overall carbon footprint of a building is lower than a brick or a concrete block. It can be sustainable if only a balance between growth and harvesting exists. While wildfires as a result of climate change are burning the Earth's wood resources at a fast pace, reproducing them does not occur as fast; simply because trees just don't grow as fast as they are cut. Only last year, about 12 million hectares of tropics was lost (Butler, 2019) which means the balance is already disrupted. UN and countries should put some limitations on the number of cuts per new plantations to ensure the balance is maintained at least nationally if not globally.

Very few companies thought of extracting new construction material from waste. For example, a German firm of Feess Eardbau has been collecting and upcycling construction materials specifically recycling concrete (Eco Africa, 2017). Oscar Andres Mendez is another example; he is an architect who, based on circular economy and a social purpose, used waste plastics and designed an affordable building system (Low Carbon City, 2015). Recycling construction material can be our way forward as the Earth does not have enough original resources left. Architects have tried to demonstrate that it is possible to build buildings from any kind of material including waste. East Sussex studio BBM designed a building for the Faculty of Art at the University of Brighton, UK, using carpet tiles clad walls that were insulated with junk, such as floppy discs and toothbrushes (Griffiths, 2014). This building was named the United Kingdom's first permanent waste house.

The most important of all is the behavior change. In North America, it is common to renovate apartment units before a new tenant comes. This means lots of good and usable construction material including wooden doors, kitchen cabinets, counter tops, kitchen sinks, wash basins, and toilets are thrown away with each move. Upcycling and reusing construction waste to build new green spaces and community farms in Enfield Regeneration project in London was one of its positive points as a sustainable urbanization project that we evaluated (Minaei et al., 2015).

POWER GENERATION AND ENERGY RESOURCE

In the times of "Climate Emergency" and worldwide attempts to solve all sorts of climate related problems to achieve Sustainable Development Goals, scientists, engineers, and policy makers are all doing their share to decrease environmental burdens while achieving social-economical sustainability. Governments talk about resilience and we, planners, promote self-sustaining urbanization, yet most of our actions push us further away from resilience or sustainability. The most important of all is our electricity consumption which gets worse day by day. With the 30% increase in energy demand by 2040 and projection of only 60% energy by renewable sources, Mwasilu and Jung (2018) predicted that fossil fuels cannot last after 2040 while ocean wave energy with little energy loss and availability in almost all coastal cities could be harnessed as one of the most reliable energy sources with potential of at least >100 kW/m average annual power density.

Industry develops and produces new products to decrease the GHG emissions and governments support those ideas and define incentives to promote a sustainable lifestyle, but in fact most of those products are heavily dependent on electricity and without it they do not even work. It is not easy to develop clean energy technologies. Innovation challenges in clean energy technology development (e.g., networked capabilities, off-grid and smart grid, carbon capture, biofuel and bioenergy, and clean energy material) are comprehensively discussed in the DTU International Energy Report (2018).

BP Statistical Review of World Energy 2019 reveals that in most countries and cities across the globe, power and electricity is still generated by burning fossil fuels, mainly coal (BP, 2019). Although some question the reliability of BP's reports, it seems that it is indeed one of the most credible sources to provide a global review as most other sources refer to it. Globally coal is the main fuel for power (electricity) generation; about 38% of the global electricity is generated by burning coal which is the same share as it was twenty years ago and this trend is on the rise (BP, 2019). Although governments aim to achieve a sustainable future, the political power of many industrial players who benefit from using coal and the number of jobs they provide force governments to support these industries (Shwartz, 2019) and they are not ready to change because it is not a financially viable option. Rare governments such as Finland have identified the issue of burning coal to generate power and heat. Helsinki recently announced 1 million dollar Helsinki Energy Challenge to work on the sustainable heating city solutions to ban coal and minimize biomass burning by 2029 and become absolute carbon neutral by 2035 (City of Helsinki, 2020). City of Toronto has partnered up with the City of Helsinki and currently is trying to shape a team for the challenge by selecting Canadian innovators and problem-solvers.

Many other countries are still blindly burning fossil fuels to generate electricity and consider it clean! Most pioneer cities now talk about the change of energy systems from fossil fuels to fully or partly renewable energies. While we still have no clear idea whether this works or not, it does not seem logical to put more pressure on the existing system by electrifying everything such as the public transport and personal vehicles before knowing the replacement solutions have actually proved their efficiency. Electrification is wrong! This should happen before we implement new incentives and bring policies such as encouraging people to buy electric vehicles or producing more electric devices. Electrification of energy services and carbon emission mitigation neither necessarily secure resources for future generations nor they prevent resource conflicts, climate change, and endangering biodiversity (Child et al., 2018).

Countries have relied on different sources to generate their energies. Turkey identified hydro, biomass, solar, wind, natural gas, coal, oil, and nuclear as alternative sources of power generation (Topcu et al., 2018). Coal, oil, and natural gas cannot be counted as alternative source of energy generation, as they belong to the fossil fuel category. In contrast, the Finnish Energy reported that only 9% of its electricity was generated from fossil fuels in 2018 and the rest was from hydro (52%), wind (10%), nuclear (21%), and biomass (6%). Solar has had 0.2% share in power generation (Kostama, 2020). Finland planned to end the coal-era by 2029, become carbon neutral[6] by 2035 and carbon negative[7] by 2050 (Sarén, 2020). Thus, some countries have achieved higher level of sustainability by using alternative and clean energy sources.

The following section will look at the main problems of each alternative source. Please note this chapter does not aim to review power generation resources, but only to bring attention to the bigger picture from a Systems Thinking point of view.

Solar Panels

Earth is running out of minerals that are critical for modern electronics and renewable technologies (Than, 2018) including photovoltaic and solar panels. At the same time our reliance on these technologies is peaking. Particularly, sources of crucial minerals to produce solar panels are ending and they have a limited life span of maximum thirty-five years with a very limited efficiency rate of maximum 23%. Also, during the process of producing solar panels carbon dioxide, sulfur dioxide, and toxic chemicals are emitted to the atmosphere that are dangerous for humans including those who work in mines and the public who lives nearby and breathes the polluted air (Thoubboron, 2018). Besides, solar panels need batteries and batteries have limited capacity to store power. They need crucial minerals too and when their lifetime is over, they become hazardous wastes which need special handling. Unfortunately, not all users and countries have that level of expertise and technology to safely recycle or discard them. Instead of looking for better options, producing more solar panels is encouraged which puts the Earth under pressure. Many villages and poor rural communities in different countries are relying on solar panels to power their mini power plants. What will happen when those solar panels or batteries stop working? Will there be another mass migrations or shrinkage of settlements? The counter argument is that new companies started recycling solar panels and electronics and new form of batteries are being invented, so it will be possible to recover a portion of the materials used. Still, we should plan to produce more renewable technologies based on very limited resources.

Wind Turbines

Studies show that wind turbines have negative environmental impacts on biodiversity of their location. People complain about their noise and reported some negative impacts on their health and well-being. They have a limited life span of 15 years and need to be replaced which is often costly. Although, we know that offshore wind turbines have a high capacity and produce a large amount of electricity (Deason, 2018).

Stockholm Exergi has been producing sustainable electricity using cooling and heat recovery method of their green data centers and dark fibers in their Stockholm Data Parks for almost twenty-five years (Stockholm Data Parks, n.d.). They could substantially decrease their carbon footprint up to 8,000 tonnes.

How should people learn to adapt with the new conditions? We are in the age of climate emergency yet many still have no clue because nothing tangible has changed in their lives. Is it the time that we think about policies and new mechanisms to train a future generation and culture that can work for their own survival?

Hydroelectricity

While many consider hydroelectricity a clean source of energy, its negative social and environmental impacts have been discussed in recent years which rules out this power generation option as a green one. The adverse consequences of disrupted ecosystem by hydro-dams goes beyond the altered wetlands and aquatic ecosystems, it has impacted food production and agricultural irrigation (Lin and Qi, 2017). Since negative environmental impacts have been observed in multiple projects, Zele˘náková et al. (2018) suggest having Environmental Impact Assessment as an early process for any hydropower proposal, even for small hydropower.

Geothermal Energy

Researchers point to the geothermal energy as a highly potential power source for countries and cities with frequent seismic and volcanic activities such as Indonesia (Pambudi, 2017). Geothermal energy does not seem to end as Earth's tectonic plates will never stop moving and by benefiting from this heated fluids and vapors under the ground and using its energy, we not only prevent more disasters, such as earthquakes and eruptions of magma, but we also tap into a never-ending source of power which is sustainable and help us move from burning fossil fuels. The value of geothermal energy industry has been predicted to rise from $4 billion in 2018 to $9 billion in 2025. In Canada, one of the first projects that invested on geothermal energy was the DEEP project in Saskatchewan. It is supposed to generate clean energy offsetting approximately 40,000 tonnes of carbon emissions and power for about 5,000 homes with the chance of benefiting from the side product of wastewater (Ratjen, 2019).

On the contrary, others consider that drilling deep holes for gas exploration can activate seismic activities and cause serious disasters such as the Lusi mud volcano (Harmon, 2009). Herman reports "We've been monitoring [The Geysers] since 1975. All the earthquakes we see there are [human] induced". Researcher conducted studies and have witnessed that earthquake occurrence and seismicity near geothermal wells have had a steady increase (Braun et al., 2016)

Our Severe Electricity Dependence in Cities

Pioneer global cities such as Toronto are going toward the concept of smart cities, electrification of utilities and transport, and application of digital smart technologies. With the technology trend of electrification claiming to reduce pollution and CO_2 emissions, and with the advancement of various forms of technologies, our dependence on electricity has been exponentially growing. This is not sustainable and in fact opposite of resilience and smartness. Our cities compete to become smart and use all sorts of technological devices that mainly work with electricity from smart streetlights to the systems that monitor them, all work if the power is on. With the emergence of smart technologies such as smart lights equipped with sensors and cameras, many cities hugely invested on these technologies and started changing their ordinary lights to smart ones to enable data collection and help moving toward a Smart City. While there are many innovative ideas to produce technologies to power streets from sustainable sources, cities are still relying on their grids. For example, solar road systems using LEDs (Barakat et al., 2016) or application of piezoelectricity in roads that is a rational option to illuminate streets (Subhalakshmi et al., 2017) without considerable GHG emissions.

All data technologies including computers, mobile phones, big data, artificial intelligence, Internet of Things (IoT), and smart city technologies work with and depend on electricity. They need crucial minerals in their chips and memories which those resources are running out too. Usage of these technologies and energy consumption is increasing on a daily basis and consequently GHG emissions. Studies have shown that even using the Internet and searching Google can contribute to carbon footprint (Quito, 2018), because although the Internet is assumed to be cloud-based, it performs on huge computers and servers which consume energy and produce a lot of heat. These servers need cooling to enable them function best and that needs power too. Norwegian government started using old mines and underground caves to store those servers in a cool and safe environment (Kim, 2019) to decrease the energy consumption for cooling. In addition, the data infrastructure and broadband networks and a variety of technological products which all release GHG emissions during their production are not often mentioned. Furthermore, in most cases these technologies are imported and shipped from other countries which again contribute to the GHG emissions. Advocates have called for a higher degree of regulation and responsibility for shipping as it is one of the main contributors to global climate change (Lister, 2015).

The dangers of power shutdowns due to the grids being hacked or natural disasters are not even discussed here, while it is a serious threat which can pause life in our modern cities and cost a lot.

Waste Management

Waste management is one of the fields that puts more environmental burden on cities. For example, our homes are stuffed with more and more electric appliances and gadgets that increase our electricity consumption. Most of them are not necessary but the consumerism culture and advertising market convince customers that they need them. This brings us to the next topic of recycling and composting. They create more waste that are often not recyclable and have to be dumped in the solid wastes.

Many countries including United Kingdom from almost a decade ago started considering all methods of energy recovery from solid recovered fuels (SRFs) including gasification, pyrolysis, and substitute fuel (NWRWMG, 2011). Recycling food wastes using an electric composter at home consuming electricity for some hours does not seem sustainable. A good example is the Sustainable Smart City of Malmo, in which food waste is turned into biogas and eventually to a fuel for public transport (BBC Studio, 2013). The whole point of composting is to decrease the amount of waste and power but these luxury electric home composters waste electricity for couple of hours to transform food waste to small amount of compost when people can produce compost in a longer period without consuming any energy on their own. On the contrary, there are sustainable examples also. Home Biogas (2020) is one of such examples that allow families to produce cooking gas and fertilizers daily by composting their food wastes. Home Biogas also designed a "bio-toilet" to help produce cooking gas and it simultaneously saves about 40,000 liters of water annually per unit which is a huge contribution to conserving water.

Smart bins are another example that many smart cities are investing on. One example is the Clean CUBE which is an IoT-based solar-powered waste compacting bin designed by Ecube Labs (2019). It benefits from a wireless ultrasonic fill-level sensor which monitors the level of trash by using WIFI to send real-time data to its network; so when the bin is full a trip is arranged for waste collection to save fuel, pollute less, and save human resources. Since it uses solar panels to generate the power needed, it can be considered as a sustainable option but if the network at the source is disconnected, this system does not work.

Food and Energy

Urban food growing at home has become a new trend. Some people who care about the environment and want to produce their own herbs at home do not know much about the real sustainability, and often buy these products thinking they help the planet. These trendy food-growing products often consume electricity and are examples of unsustainable engineering solutions. These products are also branded as "futuristic home gardening inventions" or "smart green technologies to grow food" and work in indoor environments. They are often promoted as environment-friendly options which in fact not true as they sell a sustainability story rather than a sustainable solution. What is sold it not a sustainable food-growing product but a hassle-free easy-to-do and affordable trendy product to help people feel better with their environmental choices. Most of them consume power as they provide artificial light for plants all day long. They consume clean tap water rather than green or grey water. In addition, most of these products will end up to solid waste and eventually landfills once their lifecycles are over as they are often made of non-recyclable material. From this group I can mention two examples: Lilo and "the Click and Grow" that uses nanomaterial in a so-called Smart Pots.

In case of small indoor hydroponic systems, more electricity is being used because water pumps, heaters for the fish tank to keep the water warm for fishes and lighting for the aquarium all need electricity to be able to run. EcoQube C (Freeze Lists, 2015) and Grove Garden (Grove, 2015) are only two examples of indoor hydroponic systems.

On the contrary, there are food growers that harvest rainwater and use natural light to grow fruit and vegetables indoors, a clearly more sustainable choice. There are good examples of sustainable food growing and gardening devices that use zero electricity such as Tableau from PikaPlant (2020) company which offers self-watering indoor plant pots; or ODO (Freeze Lists, 2015) which is a sustainable smart solar irrigation system that saves both energy and water.

Food keeping appliances such as fridge and freezer have become so vital in urban living that imagining life without them seems almost impossible. A Dutch designer Floris Schoonderbeek designed a prefabricated cellar to be installed underground and called it "Groundfridge"(?) which can keep vegetables, fruits, and other food relatively cool without using any electricity. This can be a sustainable and a resilient option, because it functions without using any energy consumption or GHG emission. This reminds us of the idea and concept of old ice houses in the Middle East region which kept water and food cool for a whole neighborhood.

The number of food-preparing appliances in the kitchen is on the rise, from microwave, stove, oven, toaster and sandwich maker to grinder, food processor, and slicer. All of these appliances work with electricity and mostly are plugged in all the time without even being used. They simply contribute to the vampire power[8] (Harvard Sustainability, 2018) and waste energy and release heat which in turn adds to the internal gain heat gain. Why do we have all these appliances and consume more and more energy while the mechanical versions work perfectly fine and do not waste energy at all? Perhaps we need to work on the consumerism culture and changing of behaviors now more than ever.

Most cities have switched from gas ovens to electric ovens, another choice that does not help with resilience in cities. When the power is out, none of these appliances work. There are different types of solar cookers that sustainably can provide the needed heat captured from the Sun to cook the food in tropics and areas with hot climate. One of them was invented by the MIT Professor David Wilson who tried to solve the problem of barbecues emitting carbon and polluting air; his Solar Grill stores the sun's energy for night time fuel-free grilling and has zero emissions because it does not consume any fuel (Meinhold, 2016).

CONCLUSION

Our understanding of sustainability seems to be different based on our field of studies. Throughout this chapter, I brought many examples from engineering products that were sustainable or were unsustainable but claimed to be sustainable, and that was due to not seeing the bigger picture and from the System Thinking perspective. They were the type of products and designs that did not see their impacts on our planet's ecosystem. I would recommend inventors, designers, and particularly engineers to familiarize themselves with these three useful tools: Systems Thinking, Design Thinking, and Life Cycle Assessment. If that is not the choice, I recommend collaborating with experts as well as stakeholders and end users.

There are many insightful sources on connecting innovation with engineering which also give a step-by-step guideline. Here is a list of questions based on the Design Thinking methodology that I think if inventors ask themselves in different phases of design—from idea to product—it could help them to see whether their design is truly sustainable or not.

- Does this idea solve any serious problem from urban challenges or people's life? Or is it another luxury gadget to ease the daily life and increase our carbon footprint?
- Does this product produce its own energy from renewable energies? Does it consume less energy and emit less GHG emissions compared to existing products?
- Does this product use original material and resources, or upcycled and recycled material to decrease the environmental burden of wastes?
- Can the original material and resources be found locally? If not, how far the resources should come from (shipped and flown) and what is the carbon footprint of that import?
- Is this product or its parts reusable and recyclable after its lifecycle is over?
- Does this product become a hazardous waste after its lifetime is over?
- Can this product help people save water and energy?
- How much energy and water are used during the production and how much GHG emission are emitted?

Perhaps it is the time that cities assign some limitation for energy consumption for both industries and people. We need to look for totally new solutions with fresh perspective and stop spreading the non-efficient solutions such as solar panels and wind turbines to the rest of the world.

There are serious problems in our construction industry and urbanization that the Covid-19 pandemic sheds light on and we need to think about them and find solutions. Here are some important ones which could be possible future research topics: How can we improve the closed air circulation to prevent transfer of contaminated air-born particles in a building? If central air circulation is not safe enough for building's users, what alternative technology can we rely on to conserve energy and guarantee the health of building users? How are we going to replace old solutions, such as connected exhaust fans which run through a duct in a high-rise? How are we going to ensure the closed air circulation inside a semipublic space like an airplane is clean enough that people do not get infected with air-borne diseases such as SARS or COVID-19? How can we ensure restaurants or classrooms can provide clean air for each individual if we use Plexiglass shields and protectors?

CONFLICTS OF INTEREST

In accordance with my ethical obligation as a researcher, I am reporting that I do not have a financial and/or business interest in and have not received any funding that may be affected by the research reported in the enclosed chapter.

NOTES

1. Sustainable Development Goals: The 17 Sustainable Development Goals (SDGs), which are an urgent call for action by all countries—developed and developing—in a global partnership, were introduce by United Nations in 2015. Find more here: https://sustainabledevelopment.un.org/?menu=1300.

2. GEMET defines environmental burdens as "any activity affecting the environment or any consequence of such activity which, exclusively or simultaneously, has caused or continues to cause environmental pollution, environmental risk or the use of a natural asset."

3. UHI is a phenomenon caused by lack of moisture, vegetation, and some material in urban surfaces. Buildings, roads, and most construction material retain the heat during the day and reflect it at night (Shahmohamadi et al., 2011). Because they do not allow the heat to be radiated to the sky, the temperature difference between inner city areas and the surrounding countryside sometimes reaches to 10°C or higher. It decreases the quality of air and makes the environment hotter.

4. Smart cities and sustainable smart cities: In my chapter "Place and community consciousness" (Minaei, 2017), I have comprehensively explained smart cities. In diagram 1 titled Self-Sufficient City, you can see the hierarchy of resilience,

sustainability, smartness, and self-sufficiency where I discussed some reasons for sustainability to be far more important than smartness in cities (Minaei, 2020a).

5. To read more about sustainable energy systems in buildings, I encourage you to read my former publication about sustainable architecture including ZEB, LEED, Passive House with definition, examples, criteria, and some resources (Minaei, 2020a).

6. Carbon neutral: Companies can measure the amount of carbon they emit and have two choices of either reducing the emissions or removing carbon. Some found the easiest solution which is buying carbon credits (Charlton, 2020).

7. Carbon negative: This is one step after becoming carbon neutral and companies are expected to remove more carbon dioxide from the atmosphere than they emit (Charlton, 2020).

8. Vampire power: Devices like televisions, microwaves, scanners, and printers use standby power, even when off. They continue to pull small amounts of energy, even when plugged in. Studies have found that vampire energy loads account for 5–10% of the total electricity in residential homes and for 1% of the world's carbon dioxide emissions (Harvard Sustainability, 2018).

REFERENCES

Barakat, E., N. Sinno, and M. Bernard. 2016. Intelligent street lightning system using solar panels and piezoelectric harvesters. *Journal of Electron Devices*, 23(1), pp. 1934–1939.

BBC Studio. 2013. Europe's first carbon neutral neighborhood: Smart cities—Horizons. *YouTube Video. 5:15*. Uploaded 29 July 2013. Viewed 27 March 2020. Available at: https://youtu.be/6yZYXSsWnsg

BP. 2019. BP statistical review of world energy. Viewed 19 February 2020. Available at: https://www.bp.com/content/dam/bp/business-sites/en/global/corporate/pdfs/energy-economics/statistical-review/bp-stats-review-2019-full-report.pdf

Braun, Thomas, Torsten Dahm, Frank Krüger, and Matthias Ohrnberger. 2016. Does geothermal exploitation trigger earthquakes in Tuscany? *Eos*. Available at: https://doi.org/10.1029/2016EO053197

Butler, Rhette A. 2019. The year rainforests burned. Viewed 4 March 2020. Available at: https://news.mongabay.com/2019/12/2019-the-year-rainforests-burned/

CCAO. 2019. Water facts—Worldwide water supply. *Central California Area Office, Bureau of Reclamation*. Last Modified 12 November 2019. Viewed 4 March 2020. Available at: https://www.usbr.gov/mp/arwec/water-facts-ww-water-sup.html

Charlton, Emma. 2020. What's the difference between carbon negative and carbon neutral? *World Economic Forum*. 12 March 2020. Viewed 20 May 2020. Available at: https://www.weforum.org/agenda/2020/03/what-s-the-difference-between-carbon-negative-and-carbon-neutral/

Child, Michael, Otto Koskinen, Lassi Linnanen, and Christian Breyer. 2018. Sustainability guardrails for energy scenarios of the global energy transition. *Renewable and Sustainable Energy Reviews*, 91, pp. 321–334.

Chmielewski, Frank-M., and Thomas Rötzer. 2001. Response of tree phenology to climate change across Europe. *Agricultural and Forest Meteorology*, 108(2), pp. 101–112.

City of Helsinki. 2020. Helsinki searches sustainable city heating solutions: Global one-million-euro challenge competition launches today. Last Modified 26 February 2020. Viewed 11 March 2020. Available at: https://energychallenge.hel.fi/news /press-release-helsinki-searches-sustainable-city-heating-solutions-global-one-m illion-euro

Connell, Janice, John Brazier, Alicia O'Cathain, Myfanwy Lloyd-Jones, and Suzy Paisley. 2012. Quality of life of people with mental health problems: A synthesis of qualitative research. *Health and Quality of Life Outcomes*, 10(1), p. 138.

Dales, Robert, Ling Liu, Amanda J. Wheeler, and Nicolas L. Gilbert. 2008. Quality of indoor residential air and health. *Cmaj*, 179(2), pp. 147–152.

Deason, Wesley. 2018. Comparison of 100% renewable energy system scenarios with a focus on flexibility and cost. *Renewable and Sustainable Energy Reviews*, 82, pp. 3168–3178. Available at: https://doi.org/10.1016/j.rser.2017.10.026

Eco Africa. 2017. Concrete answers to a recycling problem. 1 January 2017. Viewed 9 April 2020. Available at: https://www.dw.com/en/concrete-answers-to-a-recycl ing-problem/a-37116130

Eco Canada. 2020. Cleantech defined: A scoping study of the sector and its work-force. February 2020. Viewed 24 March 2020. Available at: http://eco.ca/research

ECube Labs. 2019.

Freeze Lists. 2015. Five futuristic home gardening inventions. *YouTube Video. 10:25*. 27 July 2015. Viewed 27 March 2020. Available at: https://youtu.be/ YaNHssp_Jqg

Gaziulusoy, Idil. 2020. System innovation for sustainability: Using systems thinking and design thinking. Viewed 15 May 2020. Available at: https://idilgaziulusoy .com/2012/06/11/system-innovation-for-sustainability-using-systems-thinking-an d-design-thinking/

GEMET. 2020. Environmental burden. Viewed 15 May 2020. Available at: https:// www.eionet.europa.eu/gemet/en/concept/15153

Ghobrial, A. 2019. Number of condos for sale drops to 10-year low. *CityNews*. 11 December 2019. Viewed 12 December 2019. Available at: https://toronto.city news.ca/video/2019/12/11/number-of-condos-for-sale-drops-to-10-year-low/

Griffiths, Alyn. 2014. Waste house by BBM is "UK's first permanent building made from rubbish". 19 June 2014. Viewed 9 March 2020. Available at: https://ww w.dezeen.com/2014/06/19/waste-house-by-bbm-architects-is-uks-first-permanent -building-made-from-rubbish/

Groundfridge. GroundFridge product. Viewed 27 March2020. Available at: https:// www.groundfridge.com/groundfridge/product/

Grove. 2015. The grove garden. *YouTube Video. 3:11*. 15 December 2015. Viewed 27 March 2020. Available at: https://youtu.be/P6qQjADFYhE

Harmon, Katherine. 2009. How does geothermal drilling trigger earthquakes? *Scientific American*. 29 June 2009. Viewed 14 May 2020. Available at: https://ww w.scientificamerican.com/article/geothermal-drilling-earthquakes/

Harvard Sustainability. 2018. Eliminate vampire power. (n.d.). Viewed 15 May 2020. Available at: https://green.harvard.edu/tools-resources/green-tip/eliminate-vam pire-power

Hobday, Richard. 2020. Coronavirus and the sun: A lesson from the 1918 influenza pandemic. *Medium.com*. 10 March 2020. Viewed 14 March 2020. Available at: https://medium.com/@ra.hobday/coronavirus-and-the-sun-a-lesson-from-the-1918 -influenza-pandemic-509151dc8065

Home Biogas. 2020. The machine that converts your waste into clean energy. Viewed 27 March 2020. Available at: https://www.homebiogas.com/

Ibrahim, S. 2013. White roof, green myth? *The Huffington Post*. 18 May 2016. Viewed November 2016. Available at: http://www.huffingtonpost.com/samir-ibra him/white-roofs-green-myth_b_2901288.html

Karlsson, Kenneth, Stefan Petrovic, and Diana Abad Hernando. 2018. Global outlook on energy technology development. In *Accelerating the clean energy revolution— Perspectives on innovation challenges: DTU international energy report 2018* (Chapter 3, pp. 21–27). Technical University of Denmark (DTU).

Kim, Jed. 2019. Norway sees a future in giant subterranean data centers. *Marketplace Tech Blogs*. 27 March. Viewed 14 May 2020. Available at: https://www.marketpl ace.org/2019/03/27/norway-sees-a-future-in-giant-subterranean-data-centers/

Kostama, Jari. 2020. Finland's and Helsinki's energy ecosystems, webinar for Helsinki challenge. Viewed 14 May 2020. Available at: Energia.fi

Lin, Zihan, and Jiaguo Qi. 2017. Hydro-dam—A nature-based solution or an ecologi cal problem: The fate of the Tonlé Sap Lake. *Environmental Research*, 158, pp. 24–32. Available at: http://dx.doi.org/10.1016/j.envres.2017.05.016

Lister, Jane. 2015. Green shipping: Governing sustainable maritime transport. *Global Policy*, 6(2), pp. 118–129. Available at: https://doi.org/10.1111/1758-5899.12180

Low Carbon City. 2015. Oscar Andres Mendez, architect and co-founder of Conceptos Plasticos. No date. Viewed 9 March 2020. Available at: https://lowcarbon.city/po rtfolio/oscar-andres-mendez-architect-and-co-founder-of-conceptos-plasticos/

Luna, G. 2015. The myth of white roofs in northern climates. *Engineering Green Buildings*. Available at: https://www.carlislesyntec.com/download.aspx?fileID =6675

Meinhold, Bridgette. 2016. Wilson solar grill stores the Sun's energy for nighttime fuel-free grilling. *Inhabitat*. 26 June 2016. Viewed 27 March 2020. Available at: https://inhabitat.com/wilson-solar-grill-stores-the-suns-energy-for-nighttime-fuel -free-grilling/

Melaas, Eli K., Jonathan A. Wang, David L. Miller, and Mark A. 2016. Friedl. Interactions between urban vegetation and surface urban heat islands: A case study in the Boston metropolitan region. *Environmental Research Letters*, 11(5), p. 054020. Available at: https://doi.org/10.1088/1748-9326/11/5/054020

Minaei, Negin. 2017. Place and community consciousness. In *Smart urban regenera tion* (pp. 68–84). London: Routledge.

Minaei, Negin. 2020a. Self-sustaining urbanization and self-sufficient cities in the era of climate change. In *Environmental management of air, water, agriculture, and energy* (pp. 175–190). London: CRC.

Minaei, Negin. 2020b. Home, windows, sunshine and your health. *ePSIch*. 2 May 2020. Viewed 15 May 2020. Available at: http://environmental-psychology.com /2020/05/02/healthy-housing-windows-sunshine-and-your-health/

Minaei, Negin, Ali Parsa, and Claudia Trillo. 2015. Enfield regeneration project (Enfield Garden). Available at: https://www.researchgate.net/project/Enfield-Regeneration-Project-Enfield-Garden

Mohajeri, Nahid, Govinda Upadhyay, Agust Gudmundsson, Dan Assouline, Jérôme Kämpf, and Jean-Louis Scartezzini. 2016. Effects of urban compactness on solar energy potential. *Renewable Energy*, 93, pp. 469–482.

Montgomery, Charles. 2013. *Happy city: Transforming our lives through urban design*. Canada: Penguin.

Mwasilu, Francis, and Jin-Woo Jung. 2018. Potential for power generation from ocean wave renewable energy source: A comprehensive review on state-of-the-art technology and future prospects. *IET Renewable Power Generation*, 13(3), pp. 363–375.

NWRWMG. 2011. The future of waste resource management. *YouTube Video. 15.* Uploaded 9 May 2011. Viewed 27 March 2020. Available at: https://youtu.be/ XGqrNi3kTLc

Pambudi, Nugroho Agung. 2018. Geothermal power generation in Indonesia, a country within the ring of fire: Current status, future development and policy. *Renewable and Sustainable Energy Reviews*, 81, pp. 2893–2901.

PikaPlant. 2020. Plants you never need to water. Viewed 27 March 2020. Available at: https://pikaplant.com/en/?v=989173909475

Quito, A. 2018. Every Google search results in CO$_2$ emissions: This real-time data viz shows how much. *Quartz*. Viewed 30 September 2019. Available at: https:/ /qz.com/1267709/every-google-search-results-in-co2-emissions-this-real-time-da taviz-shows-how-much/

Ratjen, Vanessa. 2019. Geothermal energy is taking off globally, so why not in Canada? *The Narwhal*. Last updated 2 July 2019. Viewed 11 March 2020. Available at: https://thenarwhal.ca/geothermal-energy-is-taking-off-globally-so-w hy-not-in-canada/

Sarén, Helena. 2020. Finland towards carbon neutrality, smart energy and circular economy. In *Webinar presentation, Finland's energy clusters and their supply chain ecosystems and Toronto's clean energy cluster asset map and renewable energy strategy*. 5 May 2020.

Shahmohamadi, P., A. I. Che-Ani, K. N. A. Maulud, N. M. Tawil, and N. A. G. Abdullah. 2011. The impact of anthropogenic heat on formation of urban heat island and energy consumption balance. *Urban Studies Research*. Available at: https://doi.org/10.1155/2011/497524

Shwartz, M. 2019. Stanford expert explains why we continue burning coal for energy. *Stanford Energy*. Viewed 19 February 2020. Available at: https://energy.stanford.e du/news/qa-stanford-expert-explains-why-we-continue-burning-coal-energy

Stockholm Data Parks. n.d. Green computing redefined. Viewed 11 March 2020. Available at: https://stockholmdataparks.com/benefits-of-green-computing-in-sto ckholm/#large-scale-heat-recovery

Subhalakshmi, N., A. Jayalakshmi, N. Janani, S. Induveni, and P. Dhivya Dharshini. 2017. Automatic street lights lightened by piezoelectric roads. *International Journal of Innovative Research in Science, Engineering and Technology*, 6(14). Available at: http://www.ijirset.com/upload/2017/ncmes/24_subha.PDF

Than, Ker. 2018. Critical minerals scarcity could threaten renewable energy future. *Stanford Earth*. Viewed 19 February 2020. Available at: https://earth.stanford. edu/news/critical-minerals-scarcity-could-threaten-renewable-energy-future#gs. wtmf15

Thoubboron, Kerry. 2018. Are solar panels toxic to the environment? *Energy Sage, Smarter Energy Decisions*. Viewed 4 March 2020. Available at: https://news.en ergysage.com/solar-panels-toxic-environment/

Topcu, Ilker, Füsun Ülengin, Özgür Kabak, Mine Isik, Berna Unver, and Sule Onsel Ekici. 2019. The evaluation of electricity generation resources: The case of Turkey. *Energy*, 167, pp. 417–427. Available at: https://doi.org/10.1016/j.energy.2018. 10.126

UNHABITAT. 2016a. World cities report, urbanization and development, emerging futures. *United Nations Human Settlements Programme (UN-Habitat)*. Available at: www.unhabitat.org

UNHABITAT. 2016b. Sustainable cities and communities, SDG goal11 monitoring framework. *United Nations Human Settlements Programme (UN-Habitat)*. Available at: www.unhabitat.org

Well. 2018. The international WELL building institute pbc delos living LLC, The WELL Building Standard v1, 2018. Available at: ttps://www.wellcertified.com/c ertification/v1/standard/

Zeleňáková, Martina, Rastislav Fijko, Daniel Constantin Diaconu, and Iveta Remeňáková. 2018. Environmental impact of small hydro power plant—A case study. *Environments*, 5(1), p. 12. Available at: https://doi.org/10.3390/environment s5010012

Chapter 4

Energy Efficiency and Sustainability through Wind Power for Green Hospitals

Figen Balo, Unal Yılmaz, and Lutfu S. Sua

In this study, an exemplary analysis is provided on the extent to which the current energy needs can be met by the introduction of renewable energy resources in hospitals. As a first step toward this purpose, the energy consumption of existing hospitals in a certain province was calculated using provincial health directorate. Within this stage, annual energy expenditure amounts for all sorts of energy uses such as medical devices, air conditioning systems, and lighting were determined. The second step involved the investigation of the extent to which the energy requirement of the hospitals can be obtained from the wind energy potential of the region where the hospitals are located. For this purpose, wind speed and wind direction data for one year were obtained from the Regional Directorate of State Meteorology for the predetermined province. Wind data for the latitude and longitude of the province was loaded as input to a wind energy simulation software. In the third step, the amount of wind energy available to support the generation of electricity in the province was determined annually by means of analysis. Then, turbine systems which can convert the existing wind energy to electrical energy in the most efficient way were determined. As a result, the rate of energy demand that can be met through renewable energy by installing various sizes of wind farms was determined. The concept of green hospital is a new concept and the number of applications in this field is not high. This study aims to contribute to health facility managers and policy-makers and provide a reference for future studies.

INTRODUCTION

Power has always been an ineluctable requirement for mankind. This requirement was fulfilled by traditional fossil-based energy expenditures

75

until recently. All the same, fossil-based energy utilization brought a lot of troubles. The most significant troubles linked to the fossil-based resources are climate change and environmental pollution because of greenhouse gas emission. The Earth's mean temperature is rising and this rise is causing abnormal climatic problems. Sustainable power sources have dwarn significant attention during the past decades because of these causes. In addition to the fact that fossil-based energy sources (such as coal, natural gas, and oil) will diminish in next years, the governments encourange scholars to research for optional power sources. Since the early 1970s, significant progress has been reached in various sustainable industries such as geothermal, biomass, solar, and wind energy, and so on.[1]

Hospitals provide services 24 hours a day. As a consequence, amounts of energy consumption, water consumption, and chemical waste are very high. Therefore, "green concept," which has gained importance for hospitals in recent years, has been developed to create an alternative for the heavily used resources, enable more efficient use of energy, water and materials, prevent waste, and design environment-friendly buildings.

The concept of green in health facilities has been given serious importance in developed countries. Hospitals are considered to be the second most energy-consuming commercial sector in the United States of America. In United Kingdom, the National Health Service (NHS) is the institution with the most employment and the largest expenditure. In order to minimize the damage caused to the environment, NHS carries out the green concept in all health facilities with lower carbon emission, efficient water and energy management. By analyzing most known green hospital evaluation systems (BREEAM, LEED, GREN STAR) in the world, it can be observed that energy efficiency and environmental protection are among the prominent evaluation principles. This is because one of the largest expenditure items in hospitals which provides uninterrupted services, is energy supply. Thus, the efficient and economic use of this energy is important in terms of sustainable cost for health facilities. Energy consumption in hospitals occurs in a wide range of areas such as air quality assurance devices, devices used for obtaining heat comfort, medical devices, and lighting.[2] Renewable energy sources are one of the most rational ways of meeting this intensive energy consumption in hospitals.

The wind power is among the most emerging and promising renewable sources due to its zero energy expense, abundant nature, and cleanness.[3] In addition to the technological improvements, electricity generation from wind energy has the ability to contribute favorably to the national economies. The wind turbine transforms the accessible air flow into electrical energy. Wind farms are one of the most preferred renewable energy generation plants, as they have no raw material expenses and have no expenses other than the initial investment cost and maintenance costs. Two types of aging are intro-duced to a wind energy plant. Performance loss due to relative aging, and physical tear and wear in proportion to the constantly evolving technic in the marketplace. Eventually, every farm is going to either be demolished because it is no longer valuable or replaced with novel, better performaning industry. The time between decommission and commission is called the technological lifespan.[4] The calculated turbine lifetime shows the number of years a facility is in operation. During this time, some parts may be repaired or exchanged. When the wind turbine comes to the end of its working lifetime, the rest of the parts must be removed from the site either to return the area to its former state or simply to provide space for new turbines.[5]

Content from disposed parts of discontinued wind turbines should be treated in accordance with the waste hierarchy set out by the Waste Framework Directive[6] to minimize the negatory effect on the environment. The following order of priority is proposed by the directive:

a. Reuse the parts as they are or when they are ready
b. Reduce/prevent waste (for example, by utilizing better lifetime components)
c. Material disposal (e.g., by landfill)
d. Recover the material's power (for example, by incineration)
e. Recycle the product

The preferential way is to sell either parts of facility or the whole facility for second-hand use. Nonetheless, an ordered second-hand market does not exist and technical progress is so rapid that it is hard to find use of old parts for new projects.[7] For instance, rotor blades during their lifetime are exposed to heavy tear and wear, and unless the plant is replaced before the end of its operational existence, it is therefore usually not suitable for reuse.[8] Components that are not reused should be recycled if possible for technical or economic reasons. The total recyclable waste for an entire turbine is estimated to be around 80% where most of the non-recyclable material is contained in the blades of the rotor.[8,9]

Since all major metal forms in turbines are already recyclable on a great scale, they are supposed to be less problematic than the most utilized blade materials, as no alternative for this is available at present on an industrial scale. Locally it is important to develop such an industry capable of handling these products in a sustainable way. A long-term solution could include moving to a material that is easier to recycle. The material commonly found in turbines is iron, steel, copper, aluminum, blade, and electronic element materials such as carbon or glass-reinforced PVC and plastics. In the future, the amount of waste from the world's wind energy is estimated to reach high rates compared to current levels. These projections of waste depend on the supposition that after decommissioning, no turbines or sections will be reused, which is impossible. The three diverse scenarios for development display that a well-functioning second-hand marketplace could be an effective way of reducing the amount of wind energy waste, although more research is required to draw any real results.

In every situation, the yearly amount of waste from turbines is very likely to increase substantially over the coming period of ten years.[10,11,12] This rise should be acceptable as sufficient action is taken to ensure that all resources are cared for. Failure to make so could result in landfilling or building up material to be handled in more maintainable manner, hence raising the unfavorable ecological effect of the life cycle of the wind turbine.[13]

Therefore, it is important to establish wind farms that are designed with the right simulation programs taking into account only favorable areas. Otherwise, when wind farms that are installed on a solid basis are not planned, the waste of these plants is added to resources that increase environmental pollution.

The installations that produce wind-powered electricity do not occupy much space and therefore add to ecological equilibrium security. As wind energy does not emit hazardous gas into the atmosphere, it also does not have any adverse effect on the atmosphere. The wind energy is the result of the moving air mass' kinetic energy. For practical utilization, the wind power system converts the wind's kinetic energy into electrical or mechanical power that can be utilized. Through wind electric turbines, the wind electrical production systems transform wind power into electrical power. The wind turbines produce electricity for businesses and houses and public services for sale.[14] In addition, it can also be utilized for other purposes such as pushing a sailboat, grinding grain, pumping, sawing, etc. In addition to the wind energy, table 4.1 presents the electricity generation expenses for several other sustainable power resources as well.[15]

Wind energy has a big potential globally. The wind energy's first advantage is its usability nearly everywhere on the surface of the Earth. All the same, some countries such as Turkey have usually bigger wind power potential due to their geographic parameters.

The second advantage of the wind energy is that a significant amount of power can be produced utilizing wind energy turbines. Wind parameters are needed in order to adequately plan and construct a wind enegy station. The two-characteristic Weibull dispersion function is the most adopted and effective way for the wind speed dispersion presentation. When the Weibull distribution's parameters are identifed utilizing real wind speed data, the wind power potential in a region can be evaluated by computing mean wind energy density.[16] The regional wind power sources are yet to be utilized due to the lack of itemized source evaluation statements.[17] An onsite mensuration campaign may not be essential for wind farm sites with well ascertainment to effective wind directions.[18] The primary constraints' comprehension could aid studying wind and comprehending its impact on turbines.[19] In 2007, the WINEUR project focused on obtaining accurate estimation for annual power generation. The project report did not supply a tracing methodology for wind source evaluation and the process for wind tracing emulates the process utilized in source evaluation for large-scale energy station.[20] Fields[21] proposed some key factors to be included in the course of the urban wind farm plans' technical assessment such as turbine siting, power generation, turbine definition, wind source evaluation, and the turbine's reliability and safety. They proposed atmospheric mensuration as the most accurate alternative to measure the wind source.

In this respect, it is indicated that whiledetermining the potential of wind energy, wind velocity is one of the most significant parameters. In determining the design of the wind turbine, the places where the wind turbines are to be installed and the average velocity of sustainable wind are essential parameters.

The mistake that can occur at a rate of 1% during wind velocity measurement, can trigger an error of nearly 2% in the power output.[22,23] The preliminary feasibility reports for the scheduled investments can be produced economically and technically with the assistance of the programs used for wind energy and the investors can contribute in this direction more securely and rationally. Working with the experts and the right techniques is also important in order to achieve reliable predictions in close proximity to reality. The high initial investment price is one of the most important issues in developing wind farms. Investments in positioning these facilities in inappropriate areas can result in serious financial losses with a feasibility study that can be done in the wrong way.[24] WAsP, WindPro, WindFarmer, WindSim, Windographer, Homer, and RetScreen are some of the commonly used wind power application programs for this purpose. RETScreen, Windographer, and Homer can only calculate using linear methodologies while flow-field modeling is used by WindPRO, WAsP, Openwind, and WindFarmer. On the other hand, WindSim and Meteodyn use computational fluid dynamics (CFD). There are not many comparative studies available on wind energy potential simulation. WindSim, one of the two energy calculation software programs in Torrild, Denmark, has reported 9% lower solution quality than WAsP simulation software.[25] Berge et al.[26] researched two distinct industries. The error rates achieved between the two industries by WindSim and WAsP simulation softwares were discovered to be 10–11% in one sector and 14–24% in the other. While the real value for wind energy crops in calculating the track impact is 10%, Nielsen[27] determined this value as 9 and 8%

using WindPRO and WindFarmer, respectively. Pauen[28] utilized WindSim and WAsP simulation softwares with readings taken for three specific areas at specific heights and reported comparable outcomes from the calculation of vertical wind speed.

Steinbach assessed the power density and wind speed through the vertical upgrade with the information acquired by anemometer at four different heights using both softwares. The outcomes produced with WindSim were revealed to be smaller than the real outcomes with the increasing height in terms of power density but these findings were slightly better than the ones obtained from WAsP simulation software. In the paper, it was specified that the outcomes acquired through the WindSim software by measuring over 50 m were lower in terms of wind speed than the results obtained by the actual measurements.[29] Comparative assessment of simulation softwares shows that in terms of average values, WAsP simulation software is one of the appropriate softwares. In India, WAsP simulation software incorporated into the geographic information scheme, multi-criteria assessment instruments, and aeronautical recognition coverage was performed by Mariappan and Mathew. This chapter takes into consideration not only the usability of the wind source, but also the views of societal-economic, environmental, physical, and electrical substructure, which constitute restrictions on the use of wind power.[30] Muhtasham and Khan[31] used a WAsP framework in Bangladesh to acquire the feasibility of using wind sources. Katinas et al.[32] created the wind atlas of Lithuania using a WAsP framework and documented over forty years of historical information sequence in aerology stations. The WAsP simulation software was also used by Rafiuddin and Sharma to create high-resolution wind atlas in two Fiji regions.[33] This model was also used by Carvalho et al. to assess the wind power of Portugal at two different locations. The information generated by the working the model of WRF were used as input for the simulation software of WAsP. In this research, the writers contrasted the findings of these two models with the findings of direct site measurements.[34] Ayala et al.[35] conducted both the WAsP and Urba-Wind wind map calculation code to operate the wind energy source in a complex terrestrial wind energy station and reported that both of the analytical findings underscored the actual generation. Simoes et al.[36] found that WAsP (microscale working model) and WRF-MM5 (mesoscale working model) information resources should not be jointly scheduled for urban wind qualification because these working models do not clarify the effects of urban wind circumstances and often tend to overestimate the potential of wind energy in such an environment. Hyun-Goo et al.[37] used the WRF model to enhance South Korea's wind map. Yılmaz and Balo investigated Maras, Antep, and Mardin provinces of Turkey through WAsP simulation software.[38,39,40] Balo and Sua analyzed the system equipments of wind farm for a design with optimum performance.[41,42,43,44] For this reason, it is essential to speed up the academic studies undertaken with technically accurate information in order to take advantage of wind energy, which is one of the most significant renewable energy sources, and to carry out the required feasibility studies in areas with sustainable wind potential in the light of these research activities. In specific terms, it has become necessary to use environment-friendly, climate-friendly, sustainable, renewable, and clean energy technologies.

In this chapter, Diyarbakir Province's wind power potential was investigated. Diyarbakir province is located in the middle of Southeastern Anatolia region. Within the scope of this research, the hourly wind information at this province is obtained at a height of 10 meters for one year from the State Meteorology General Directorate. This data was uploaded to the WAsP software. Based on the assessment of wind energy, steady wind direction, wind speed, wind power, and the wind turbines' power capacity were determined for this province. Subsequently, electricity consumption of all public hospitals within the province was obtained from the provincial health directorate. As a result, the amount of electricity that can be obtained with the assistance of a wind farm designed in the nearest area with sufficient potential to public hospitals in Diyarbakır was determined. This way, the electrical energy amount obtained from the wind farm was investigated for contribution to the energy demand of all public hospitals in the region.

WIND ATLAS ANALYSIS AND APPLICATION PROGRAM SIMULATION SOFTWARE

In 1987, the WAsP simulation software was developed in Riso Meteorology Laboratory of the Danish Meteorological Organization and improved by the Department of Wind Energy at Denmark Technical University to obtain the statistics required to analyze potential of the wind energy. For representation of the wind climatic statistics' vertical-horizontal extrapolation, the WAsP simulation software uses overall list of the models[45] and the prediction of wind sources and climatic wind.[46] The WAsP simulation software is a lineal digital working model which depends on the physical foundations of the flows in the atmospheric boundary layer. Through a few comparisons, this simulation software has been confirmed between modeled-measured wind farm productions and wind statistics.[47] The simulation software is competent of explaining wind flow over various lands, at private points and near sheltering impediments. Today it is used for regional reasons to determine the potential of wind energy.[48] The WAsP simulation software has been used in wind energy aerology for more than twenty-five years and has become the flow design technology standard for evaluating wind sources and sitting wind farms-wind turbines. It can be used conveniently and independently as an "internal computing motor" for energy output calculations and wind source evaluation within the WindPRO modules suite. The WAsP simulation software is a software for horizontal and vertical extrapolation of the wind climate statistics. It is a software that can perform an assessment using wind speed information from meteorological stations and analyze information by modeling wind direction on the ground. This simulation software conducts the study on the assumption that the studied region's wind velocity information has a distribution consistent with the Weibull 2-parameter distribution. The WAsP simulation software calculates the statistics of the national wind atlas by assessing four distinct input information in its submodels.[25] Required inputs for this simulation include hourly wind data (wind velocity and direction), topography of the region where the measuring station is situated, area information on roughness, and barrier information about the surrounding setting around the wind measuring station. The WAsP simulation software consists of four loader documents from Microsoft, the WAsP Climate Analyst, the WAsP Map Editor, and the WAsP software. Also establishing the WAsP Turbine Editor is the WAsP placement program. Processes conducted in these blocks include an assessment of the time-ordered information, estimation of the wind system, generation of wind atlas information, and assessment of the potential of wind power in relation to the calculation of the complete wind farm energy output to be established.[49] It includes some physical models to identify wind flow near shelter obstacles and across various terrains. Documentation is included as records of internet assistance. The WAsP simulation software runs on Windows, XP, 7, and Vista.

Traditionally, wind farm computations and wind source evaluations were dependent upon wind data gauged at or near wind farm location by means of wind map methodology.[50] At the wind farm site, wind readings aretaken generally every 10 minutes throughout the year. The data logging system converts these raw site wind data into calibrated wind data using calibration expressions for each instrument. The quality and integrity of the calibrated wind information is then evaluated through time-series visual inspection and data analysis. Values obtained from other comparable or redundant sensors may be replaced by missing information. The goal is to obtain the most precise, reliable, and complete information set. Next, this information set should be viewed in the context of the long-term wind climate of the site and an adapted information set that represents the long-term climate should be created. When establishing an information set representing the long-term climate at the site, this can be used to calculate the wind environment stats such as wind velocity and wind direction distributions, as well as mean values, standard deviations and other statistics. The final stage in the wind resource assessment process is to predict the long-term wind conditions at the forecast locations.

The aerological modelings are utilized to compute the generalized wind clime science using the analysis and the measured data. In the inverse operation, the wind map data implementation of the wind clime at any private place may be computed from the universalized wind clime science. In the computations, the wind farm evaluation tool entails basic tools to help. The WAsP simulation software is dependent upon two basic assumptions: initially, the universalized wind clime is supposed to be almost the same at the aerological station and estimated places (wind turbines) and in the second stage, the bygone (historical wind data) is supposed to be characteristic of the next one (over the twenty-year lifetime of the wind turbines). The credibility of any obtained WAsP simulation software estimate is based on the degree to which these two assumptions hold.

- Place wind clime = Place wind data ± [longtime extrapolation impacts]

  Utilizing a longtime extrapolation process, the place wind data are adjusted and referenced with respect to the longtime clime science of the field.
- Reference annual energy production = Wind clime at center peak plus (energy curve)

  The reference annual energy production is computed utilizing the predicted wind clime at center peak at the pole place and the site—private wind turbine energy curve. All the same, heaps of times this stage is exceeded and the gross annual energy production is immediately computed.
- Gross annual energy production = Reference annual energy production ± (site impacts)

  Utilizing a flow modeling, the monitored wind clime at the mast place is converted to the estimated wind climatic at the wind farm's wind turbine places.
- Potency annual energy production = Gross annual energy production − (wake losses)

  At each of turbines place, utilizing a wake modeling, the wake losses are subtracted and predicted from the gross annual energy production. This corresponding to the Wasp "net annual energy production."

  Net annual energy production = Potency annual energy production − (techn ic losses)

- In the wind farm, the extra operational (technical) losses are next subtracted and predicted from the potency annual energy production to obtain the net annual energy production value (P 50) at the widespread coupling's point (PCC).

    P 90 annual energy production = P 50 annual energy production – [indefiniteness prediction] × 1.282.
- The whole power efficiency process's aggregate indefiniteness is predicted and the net annual energy production is set to obtain a net value corresponding to exceedance's particular probability.

In short, there are several kinds of "prediction" or "estimation" at play here: first, we estimate what the wind climate has been like in the past at our site mast, by referencing our observations to a suitable long-term data set. Second, we try to predict what the wind climate has been like for our wind turbine.

Weibull Distribution Function Used by WAsP Simulation Software

The wind speed distribution modeling is of big significance for the efficiency of the wind power transformation system and the evaluation of wind energy potential. WAsP simulation software was utilized within the scope of this study. The Weibull dispersion function provides feasibility to define the wind speeds' periodicity distribution over a certain duration. All the same, for the analysis of a site's wind potency, this dispersion is utilized for a time period more than one year. It is characterized by using equation (4.1)[49,51]:

$$f(V) = \frac{k}{c}\left(\frac{v}{c}\right)^{k-1} \exp\left[-\left(\frac{v}{c}\right)^{k}\right] \tag{4.1}$$

where, the Weibull shape characteristic (dimensionless) is k, the wind speed is v, and the Weibull scale characteristic is c (m/s). For a specific area, the determination of these parameters considers the wind regime's modeling. The F(v) cumulative dispersion function serves as the duration's (or possibility's) fraction where the wind speed is equal to v or less. The Weibull distribution's cumulative possibility function is determined using equation 4.2.[52,53]

$$F(v) = 1 - \exp\left[-\left(\frac{v}{c}\right)^{k}\right] \tag{4.2}$$

To forecast Weibull c and k characteristics, we utilize the system solution by sequential iterations of the modified maximal possibility optimization methodology under WAsP simulation. The Weibull characteristics are forecasted by simulation software utilizing equations (4.3) and (4.4) when wind speed information is present in the periodicity dispersion format.

$$k = \left(\sigma / \overline{v} \right)^{-1.086} \left(1 \leq k \leq 10 \right) \tag{4.3}$$

$$c = \frac{\overline{v}}{\Gamma \left(1 + 1/k \right)} \tag{4.4}$$

where, gamma function is shown with Γ. The standard declination σ and the average wind speed $\overline{v}$ can be expressed as shown in equations (4.5) and (4.6):

$$\overline{v} = \frac{1}{n} \left(\sum_{i=1}^{n} v_i \right) \tag{4.5}$$

$$\sigma = \left[\frac{1}{n-1} \right] \sum_{i=1}^{n} \left(v_i - \overline{v} \right)^2 \tag{4.6}$$

where, v_i is nonzero wind speed, n is the nonzero wind speed data number, f is the periodicity for which the wind speed diminishs, F ($v > 0$) is the possibility of the wind speed to be equal to zero or greater, and n is the period number.

DIYARBAKIR PROVINCE IN TERMS OF WORKING PARAMETERS USED IN WASP SIMULATION SOFTWARE

The Geographical Characteristics

The province of Diyarbakir has an area of 15,355 km² and is located 660 meters above sea level. It is between 40° 37″ and 41° 20″ east longitudes and 37° 30″ and 38° 43″ north latitudes. The surface shapes are quite plain. It is surrounded by heights. The center is a pit basin. The axis of this pit area, called the Diyarbakir basin, forms the wide Tigris Valley in the west-east direction. It is surrounded by the Taurus Mountains arc from the north. These mountains divide the Eastern Anatolia region and Southeast Anatolia. Its altitude reaches 1.957 meters at its peak Kolubaba. A harsh land climate prevails in Diyarbakir.[54]

Total Wind Direction: Force and Climate
Data for Twelve Years in the Province

The current installed power of the region's power plant is 2,253 MW. Existing twelve power plants generate approximately 7.409 GW of electricity annually. There are seven licensed and five unlicensed electricity generation companies in the province. The consumption rate to national average is 2.9%.[55] Although the wind in the region compared to the country in overall is small, still a potential exists.

A harsh land climate prevails in Diyarbakir. The most available wind speed values are obtained in winter and fall months according to the average climate data. Summers are very hot, but winters are not as cold as they are in Eastern Anatolia. The main reason for this is that the Southeast Taurus Mountains cut the cold winds coming from the north. The temperatures are highest on average in July, at around 29.7°C. At 2.2°C on average, January is the coldest month of the year. The least amount of rainfall occurs in August. The average in this month is 0 mm. In January, the precipitation reaches its peak, with an average of 79 mm. Between the driest and wettest months, the difference in precipitation is 79 mm. The variation in temperatures throughout the year is 27.5°C.[54]

Economic wind energy investment requires a wind speed of 7 m/s and a 35% capacity factor. Areas can be evaluated as wind power installable areas by considering the available wind speed values.

Wind power plant capacity to be installed in the province is given in the following table. It can be seen that it is possible to establish a power plant with an installed capacity of 635.04 MW in an area of 127.01 km² which can be evaluated economically (table 4.2). Even this value is a very serious potential. This potential can meet all the electricity needs of the province.[54,56]

In this study, the wind speed and wind direction data measured at 10-meter height in recent years in Diyarbakir are obtained from the State Meteorology General Directorate. These values are used as input to the WAsP simulation software to determine the potential of wind power in this region.

The Electricity Energy Consumption of
Hospitals in Diyarbakir Province

Hospitals are complicated structures with distinctive energy needs that exceed many other kinds of structures. They operate on a 7/24 basis to serve many people. Medical demands require rigorous control of the heat setting and parameters of indoor air. The use of energy is further increased by specialized medical equipment, sterilization, laundries and food preparation.

Given the hospitals' vital role in offering health care to communities, hospital energy demand management has long been deemed of little significance. Management policies for facilities tended to focus on the reliable running of construction and building services (including emergencies such as black outs) and compliance with rigorous health and safety and other clinical requirements. But elevated energy expenses, as well as climate change legislation, are increasingly prompting efforts to decrease the use of hospital energy in some nations as particular as industry carbon objectives. In Diyarbakir province, hospitals' electricity energy consumption can be generally divided into the categories as follows:

• Various loads—fans, computers, washing machine, kettle, microwave, etc.
• Air conditioner—inverter split units
• Medical devices—compressor, autoclaves, dental chairs, and so on.
• Lighting—LED armatures
• Refrigeration—refrigerators

Twenty-seven public hospitals exist at the disposal of the Diyarbakir provincial health directorate.

In table 4.3, these hospital's names and electrical energy consumptions are given for the period of June 2018–June 2019. For the twenty-seven public hospitals, multiper, active, point, and night values are determined throughout the year. In this table, the determined values are calculated as mentioned below:

Active = Active First Index - Active End Index
The Point = Point First Index - Point End Index
Night = Night First Index - Night End Index

Total electrical energy consumption between June 2018 and June 2019 is determined as 36,428,442.86 kWh. The highest and lowest energy consumption values are obtained from *Gazi Yasargil Training And Research Hospital* (11,905,791.3 kWh) and *Selahaddin Eyyubi Annex 2* (8,644.14 kWh), respectively.

The municipal electricity provider's supply between June 2018 and June 2019 is subject to disruption with interruptions ranging from a few hours to an entire day. Because Diyarbakir is a mountain town with elevated wind speeds in winter, we can expect the biggest contribution of wind energy when they are most useful. The Wind Turbine energy production will be adequate in periods of excellent wind resource to satisfy the portion of the hospital's energy demands as generation becomes available. Theoretically, an embedded system should investigate this renewable power scheme to ensure 100% accessibility.

DETERMINING THE POTENTIAL FOR WIND
ENERGY WITH WASP SIMULATION SOFTWARE

Although the province is growing quickly, the industry has only lately begun to evolve, thus increasing the need for electricity. Accordingly, numerous alternative techniques for solving the increasing energy demand have been suggested. Wind power is remarkably competitive compared to other techniques of generating electricity. Both meteorological and economic parameters are required to be evaluated simultaneously in order to create an efficient wind power plant. This also demonstrates the fact that planning a wind farm requires more than one discipline to be involved in the design phase.

Wind power is becoming the least costly and fastest-growing viable energy source. The environmental benefits consolidated with the significant economic advantages consisted of an affirmative prognosis for continuously growing wind power. The milestones that wind power has reaced in recent years, such as a drastic reduction in energy expenditure, the power to deliver a significant part of electrical demand and enhancing its grid inclusion credibility, have resulted in an average increase of nearly 25% in wind energy usage. In essence, greater turbulence and lower wind speeds due to the existence of obstructions complicate things and describe the wind circumstances in a certain region. Potential assessment in terms of wind continuity with precise parameters of a region is highly essential for the feasibility of the planned wind farm.

In this study, the evaluability of wind energy, one of the most sustainable energy sources, for public hospitals in the province of Diyarbakir was investigated. The proportion of the electrical energy needs of the public hospitals in the region was researched. For this purpose, firstly the data of the wind energy of the province in recent years was obtained from the meteorology general directorate. Then the obtained data was used as input to the WAsP simulation software. The amount of electrical energy that can be obtained by benefiting from the wind potential of the region and the turbines to be used to provide this energy have been determined. Then, the amount of electrical energy consumption of public hospitals in Diyarbakir for the last one year was obtained from the provincial health directorate. The amount of electrical energy calculated by the WAsP simulation software for the province was investigated to determine the extent that the regional public hospitals could meet the electrical energy.

In this study, the meteorology station is located at coordinates (37.53753, 40.10769) by Global Mapper.[23] Diyarbakir City's roughness and orography maps were prepared using the WAsP Map Editor software. WAsP Map Editor helps transform images to digital WAsP maps representing roughness and orography using scale calibration. The mountain axis is almost perpendicular to the current wind direction, making it an ideal location for wind farm projects.[57] The numerous wind power effects display the South-West winds path. The source map field shows the wind situation and roughness (by isohyped dem map) of all towns in Diyarbakir. The above sea level ranges from 0 to 675 meters.

The area where the anemometer was located at the center of Diyarbakır was drawn as 100-meter spaced and the area reached 700-meter altitude while the highest altitude in the center of Diyarbakır was 2,230 meters. Wind data for one year at a height of 10 meters was used as input to the WAsP simulation software-. The observed wind climate anemometer data were collected at 37,53°N 40,10°E.

For an application location (turbines), the estimated wind climate includes the WAsP application's results, the generation of power and the predicted wind climate. When the results are computed by the WAsP model, the expected window shows the graph and the expected wind climate. The following outcomes are achieved for the annual information, including wind speed and direction gathered at the MET tower, after taking into consideration the topography variables. Using the topographic map along with the place and the chosen features of the wind turbine, these findings are used for projection for the wind farm. The prevailing wind direction is from north to west. All obstacles winding up in this direction will significantly contribute to wake-up losses at the wind farm.

As a result, the general wind direction of the anemometer was modeled in northwest orbit and the average wind speed and the wind power passing through the unit area was found to be 3.32 m/s and 57 W/m^2, respectively. The power density value for standard air density is calculated to be 1,117 kg/m^3. The mean AEP value is obtained as 548,172 MWh.

The energy window demonstrates the graph and rise depend on the region's generalized wind atlas (wind climate) for the predicted energy density or yearly energy generation at the turbine area. The rise in wind energy density determines the contributions from the various industries to complete energy density.

The P and AEP show the sector-specific contributions to total power density and total power output, taking into account the frequency of wind occurrence. All energy output provided at the bottom of the grid is calculated from the amount of industry outputs. This may not be entirely the same as the total wind dispersion's production depending on the Weibull A- and k characteristics; thus, the difference should be small in most cases.

Analysis of the wind direction is a significant part of the source evaluation as it shows the prevailing wind direction from which to harness the energy. A dominant industry is preferred to a supply of distributed power.[58] The wind velocities are extremely variant in the direction of different locations;

thus, the area covered by the source grid is within the predictability limits of WAsP. The radar map of the wind direction indicates that the prevailing wind direction studied over the year is generally between 285 and 315 degrees. In other definition, the distribution of the wind directions shows that the northeast of Diyarbakir and more precisely the second sector which has an inclination in comparison with the ranging geometric angle enters (30°–60°) north-west has the wind power with highest performance. These range corresponds to about 80% of the tower's overall wind flow. This region shows better potential due to the regional alignment prevailing north-west winds. This is primarly due to the dense forest fields in these regions, which a higher roughness value is reflected on the WAsP Map Editor.

As shown in table 4.4, the temperature value was about 14.340°C, while the relative humidity was 45.50808.

Precise estimates of the capacity for wind power depend on wind patterns knowledge at the wind turbine hub height. On the other hand, due to mast expense and higher measurement, wind measurements are usually applied below wind turbine hub heights. It is necessary to know the wind speed vertical profile to make a reliable prediction of the wind turbine output. From the European Wind Atlas, the Wind Atlas Analysis and Application simulation software is based on methodology. The program uses wind speeds estimated at only one height of measurement as input data.

Annual energy output [GWh] determined by the WAsP for highest and lowest air density turbines and designed wind facility by utilizing with power curves are tailored to individual turbine sites for selected reference air densities. Usually, the energy yield (MWh /A) is calculated to be more or less equal to the density of wind power (W/m^2). The difference in power density is only due to variability in the distribution of wind speed. The color scale shows the density of wind power in Wm^{-2}. For the measurement of the power density, the air density is from the simulation of the mesoscale model. Nevertheless, the wind power density and therefore the AEP also depend on the air density at hub height, but very little attention has been paid to the estimation of air density in the literature.

The wind resource grid displayed as a map over a chosen region is a helpful characteristic in WAsP. This can be utilized to identify AEP sites that are possibly large. In this way, an estimation of a wind turbine's average AEP (annual energy production) can be obtained by supplying the wind turbine's power curve through WAsP.

wind turbine AEP (annual energy production) —> power curve + estimated wind climate

If a range of wind turbines are built near eachother, they are called a wind farm. The turbines communicate with each other in a wind farm. It is also possible to calculate the wind farm output using WAsP.

Given the thrust coefficient curve of the wind turbine(s) and the wind farm layout, WAsP can estimate the wake losses for each turbine in the farm and thereby the net annual energy production of each wind turbine and of the entire farm, that is, the gross production minus the wake losses. Mixed wind farms, with different turbine types and hub heights, are allowed.

Net AEP of whole wind farm—> wake losses + annual energy production of wind turbines

Using the RIX (ruggedness index) is part of WAsP's advanced application. As an objective measure of the extent of steep slopes in an area, WAsP may obtain a definition of RIX. Also, how to estimate prediction errors using the difference in RIX values between the meteorology station (on which the wind atlas is based) and the site(s) of the wind turbine. This could be used to allow correction of output to conform to the final forecast of WAsP. First of all, the RIX value is a metric by which you can determine whether WAsP operates outside or within its performance envelope, whether errors of prediction can be predicted, and the sign and estimated magnitude of such errors.

In the terrain, the WAsP simulation program can predict the average wind climate at all the sites in a wind farm and anywhere. It is the industry standard for wind source siting, energy yield, and calculation assessment for wind farms and wind turbines. The WAsP simulation program can be utilized for areas situated in all types of land all around the world. The meteorology station data saved for a period of one year is utilized to project direction and wind speed at each of the turbines location. For an entire wind farm, the WAsP simulation program can predict the single turbine site's energy generation. It considers layout, wake losses, and numerous other elements. At a turbine location, the wind energy density is based on the cubed wind speed, so there is generally a strong focus on modeling and measuring the wind speed periodicity dispersion with correctness that is as high as possible. If wind direction and speed have been gauged at two or more meteorological masts within the wind station location or within the zone in which the regional wind climate is supposed to be similar, it is feasible to ratify to what extent WAsP is able to model the wind speed differences across the wind station location. This knowledge may be utilized to set the atmospherical stability adjusting and the land definitions to fix the WAsP model of the wind station location.

Wind farm developers also differ for the prediction of losses caused by wakings in a multi-turbine project. This study uses WAsP software, which is widely used in the analysis of awakeness in a wind farm with field assessment tools. Taking into account the geographical and ground surface conditions, the change in wind speed profiles can be estimated with the expected wake propagation intensity. This estimate minimizes awakeness losses while allowing turbines to "micro-sit" or individual placement in a wind farm. The most important factors affecting the wake losses are the incoming wind speed, wind direction and turbine position. Naturally, when a table is created with a single turbine and for speed not running this turbine, the wake loss is shown with the value zero.

CONCLUSIONS AND FURTHER OUTLOOK

This study was carried out as a preliminary study to determine whether the potential that can be obtained from wind energy in the region will meet the energy needs of existing public hospitals in the region. If sufficient renewable energy potential is identified in the region, a more detailed feasibility study should be carried out before implementation. Along with the results obtained from such studies and a detailed feasibility study, it is the subject of a new study planned to carry out wind energy application to the power plants. In this chapter, as a result of the simulation study, the turbine brand and power suitable for the topography determined by the WAsP program are presented as suggestions.

Average wind speed may vary from year to year. Due to the variability of the wind speed, the energy to be obtained from the wind energy potential is more than the energy calculated from the annual average speed value. Due to the increase in the size of the turbines that reach maximum power at lower speeds, the energy produced increases more than the predicted installed power capacity. Therefore, in the calculation of the amount of electrical energy that can be produced with wind turbines in a certain region, the wind speed frequency distribution calculated by the observed distribution or Weibull distribution is used rather than the annual average wind speed. That is, the amount of energy produced by the turbine depends on the wind speed distribution. Since the density of the air is small, the energy to be obtained from the wind depends on the wind speed. Wind speed increases with altitude and wind power increases with the cube of its speed. Depending on the frequency distribution, power density differences can be doubled in different places with the same average wind speed. This is due to the cube multiplier. The turbines usually start at 4–5 m/s wind speed and reach maximum power at a wind speed of about 15 m/s. At very high speeds, like the storm speed of 25 m/s, wind turbines turn off because the turbines stop working at very high or very low wind speeds. Wind energy conversion systems depend on the energy supplied by the wind, its power, and the number of hours of blow. The wind power is the power per unit surface perpendicular to the air flow. According to topographic conditions, the specific power at a height of 50 meters from the ground can be less than 50 W/m^2 when the speed is less than 3.5 m/s or more than 1,800 W/m^2 when the speed is greater than 11.5 m/s. Turbines that are positioned higher or are made up of new-generation materials (e.g., carbon blades) as well as larger diameter turbines are expected to provide solutions in the near future even if they do not currently offer a complete solution, because the combination of larger rotors and high towers can enable the turbine to receive higher wind speeds and generate more energy. More cost-effective and higher towers can be created with high technology wings, advanced load and controls. For example, research on new turbines, the rotor of which can be adapted to various conditions with customized carbon blades, depending on specific customer and field requirements, is available and is constantly being developed.

According to the results obtained, this study does not claim that there might be a feasible farm design in the region. The simulations only provide suggestions on the turbines most suitable for the specific topography only among the turbines available in the market. However, this study is presented as a reference study to indicate the extent to which it can be supported with renewable wind energy according to the current potential of the region in terms of the establishment of a potential farm for the region when new-generation wind turbines that are developing rapidly can be used.

This chapter examines the features and potential of wind energy in a predetermined province, using the average wind velocity information of the recent years obtained from the General Meteorology Directorate. The wind energy potential in the province was simulated using WAsP (Wind Atlas Analysis and Wind Energy Application Simulation Software), which is one of the common research topics of the recent years. Based on post-processed wind information, WAsP determined the region's climatology, local orography, and ruggedness. Then, the wind power potential of the province was assessed and it is shown that the zone is appropriate for wind power generation utilizing appropriate wind turbines of scale. The dominant direction of wind speed was determined to be southeast. This research involves the investigation of the extent to which the energy requirement of the hospitals can be obtained from the wind energy potential of the region where the hospitals are located. While a feasible wind energy farm in Diyarbakir province will generate electricity of 548,172.00 MWh/year, the values obtained are higher than all the public hospitals in the province working full load hours. In addition, the predictable wind-generated electricity prices are smaller than the local electricity tariff.

For businesses designing wind farms, this research is essential in terms of being a reference for pre-feasibility and allowing the comparison of the outcomes acquired with the simulation and distinct techniques at green hospital design. It is also intended to increase awareness that the zone can contribute to the regional and national economy through the use of the wind-related form of renewable energy available for green hospital designs in the region, as it is determined that the region has wind potential.

NOMENCLATURE

c Weibull scale characteristic (m/s)
f periodicity
k Weibull shape characteristic
n nonzero wind speed data number (m/s)
T tempreture (°C)
v wind speed (m/s)
$\bar{v}$ average wind speed (m/s)
v_i nonzero wind speed (m/s)
Γ gamma function
σ standard declination

NOTES

1. Delarue ED, Luickx PJ, D'haeseleer WD. The actual effect of wind power on overall electricity generation costs and CO_2 emissions. *Energy Conversion Management.* 2009;50:1450–1456.

2. Facilities management—Energy efficiency. http://www.hercenter.org/facilit iesandgrounds/energy.php.

3. Shahizare B, Nik-Ghazali N, Chong WT, Tabatabaeikia S, Izadyar N, Esmaeilzadeh A. Novel investigation of the different omni-direction-guidevane angles effects on the urban vertical axis wind turbine outputmardin power via three-dimensional numerical simulation. *Energy Conversion Management.* 2016;117:206–217.

4. Andersen N. Wind turbine end-of-life: Characterisation of waste material. Master thesis. University of Gävle, 2015.

5. L. Aldén. Disassembly of wind turbines and finishing of the site. Swedish Energy Agency, Visby, 2013.

6. EU. Directive 2008/98/EC on waste: European Commission. 3 March 2015. [Online]. Available: http://ec.europa.eu/environment/waste/framework/. [Accessed 22 April 2015].

7. Vindenergi S. Wind turbines: Mapping of activities and costs of dismantling, restoration of space and recycling. Stockholm, 2009.

8. Cherrington R, Goodship V, Meredith J, Wood B, Coles S, Vuillaume A, Feito-Boirac A, Spee F, Kirwan K. Producer responsibility: Defining the incentive for recycling composite wind turbine blades in Europe. *Energy Policy.* 2012;47:13–21.

9. Ortegon K, Nies LF, Sutherland JW. Preparing for end of service life of wind turbines. *Elsevier: Journal of Cleaner Production.* 2012;39:191–199.

10. Akyıldız NA, Polat H. Perception of space modernity and postmodernity transformation process. *International Journal of Scientific and Technological Research.* 2018; 4(10): 627–634.

11. Polat H, Arıoğlu N. Quality control analysis during housing construction process. *International Journal of Scientific and Technological Research*. 2018; 4(10): 440–447.

12. Ergin S. The role of climatic factors in traditional residential architecture in Diyarbakır Suriçi region, Dicle University Journal of Engineering Faculty, 2017.

13. Ergin S. Use of materials and architectural features in the rural architecture of Diyarbakır Province, Anadolu 2nd International Applied Sciences Congress, 2019.

14. AWEA. What is wind energy? American Wind Energy Association (AWEA), 2008. http://www.awea.org/faq/wwt_basics.html.

15. REL. Renewable Energy Law, The Ministry of Energy and Natural Resources, 2010.

16. Senel B, Senel M, Bilir L. Role of wind power in the energy policy of Turkey. *Energy Technology Policy*. 2014;1:123–130.

17. Toja-Silva F. On roof geometry for urban wind energy exploitation in high-rise buildings. *Computation*. 2015;3(2):299.

18. Smith J. Built-environment Wind Turbine Roadmap. National Renewable Energy Laboratory, 2012.

19. Simoes T, Estanqueiro A. A new methodology for urban wind resource assessment. *Renewable Energy*. 2016;89:598–605.

20. Kesby JE, Bradney DR, Clausen PD. Determining diffuser augmented wind turbine performance using a combined CFD/BEM method. *Journal of Physics: Conference Series*. 2011;753:1–10, 6.

21. Fields J. Deployment of wind turbines in the built environment: Risks, lessons, and recommended practices. National Renewable Energy Laboratory, 2016.

22. Sagbansua L, Balo F. Decision-making model development in increasing wind farm energy efficiency. *Renewable Energy*. 2017,109:354–362.

23. Sagbansua L, Balo F. Comparative assessment of wind turbine alternatives for wind farms. *International Journal of Engineering Science and Computing*. 2016;6(12):3661–3666.

24. Güzel S. Rüzgar enerjisi potansiyel hesaplamasinda kullanilan bilgisayar programlarinin karşılaştırılması. Yüksek Lisans tezi, ITÜ, 2014.

25. Hocaoğlu FO, Kurban M, Filik ÜB. WAsP Yazılımı İle Rüzgar Potansiyeli Analizi Ve Uygulama.

26. Berge E, Gravdahl AR, Schelling J, Tallhaug L, Undheim O. Wind in complex terrain: A comparison of WAsP and two CFD-models.

27. Nielsen P. Comparing WindPRO and Windfarmer wake loss calculation, 2002.

28. Pauen R. Vertical profiles with WindSim and WAsP comparing several cases.

29. Steinbach E. Comparison of measured wind profiles with WindSim calculations, 2007.

30. Mathew SA, Mariappan N. Wind resource land mapping using ArcGIS, WAsP and multi criteria decision analysis (MCDA). *Energy Procedia*. 2014;52:666–675.

31. Khan KS, Muhtasham H. A pre-feasibility study of wind resources in Kutubdia Island, Bangladesh. *Renewable Energy*. 2006;31:2329–2341. http://dx.doi.org/10.1016/j.renene.2006.02.011.

32. Katinas V, Sankauskas D, Markevičius A, Eugenijus P. Investigation of the wind energy characteristics and power generation in Lithuania. *Renewable Energy.* 2014;66:299–304. http://dx.doi.org/10.1016/j.renene.2013.12.013.

33. Sharma K, Rafiuddin AM. Wind energy resource assessment for the Fiji Islands: Kadavu Island and Suva peninsula. *Renewable Energy.* 2016;89:168–180. http://dx.doi.org/10.1016/j.renene.2015.12.014.

34. Carvalho D, Rocha A, Silva Santos C, Pereira R. Wind resource modelling in complex terrain using different mesoscale–microscale coupling techniques. *Applied Energy.* 2013;108:493–504. http://dx.doi.org/10.1016/j.apenergy.2013.03.074.

35. Ayala M, Maldonado J, Paccha E, Riba C. Wind power resource assessment in complex terrain: Villonaco case-study using computational fluid dynamics analysis. In: *3rd International Conference on Energy and Environment Research, ICEER 2016, Energy Procedia,* Barcelona, Spain, 2016.

36. Simoes T, Costa PA, Estanqueiro AI. A first methodology for wind energy resource assessment in urbanised areas in Portugal. In: *European Wind Energy Conference Proceedings,* Marseille, France, 2009.

37. Hyun-Goo K, Young-Heack K, Hyo-Jung H, Chang-Yeol Y. Evaluation of inland wind resource potential of South Korea according to Environmental Conservation Value Assessment. *Energy Procedia.* 2014;57:773–781. http://dx.doi .org/10.1016/j.egypro.2014.10.285.

38. Yılmaz Ü, Balo F. Simulation analysis of wind energy powered electrical energy production for Gümüşhane Province. In: *International Congress on Sustainable Agriculture and Technology (INCSAT),* Gazıantep, 1–3 April 2019.

39. Yılmaz Ü, Balo F. The simulation supported analysis of wind energy potential ın Van Province (Rüzgâr Enerjisi Potansiyelinin Van İli İçin Simülasyon Destekli Analizi). In: *International Symposium on Advanced Engineering Technologies (ISADET 2019),* Kahramanmaraş, 2–4 May 2019.

40. Yılmaz U, Balo F, Sua LS. Simulation framework for wind energy attributes with WAsP. Procedia Computer Science Elsevier ISSN: 1877-0509.

41. Sagbansua L, Balo F. Comprehensive decision-making for evaluating wind turbines. *International Journal of Science, Environment and Technology.* 2017;6(1). ISSN 2278-3687.

42. Şağbanşua L, Balo F. Multi-criteria decision making for 1.5 MW wind turbine selection. In: *8th 2016 2nd International Conference on Renewable Energy and Development, (ICRED 2016),* Kitakyushu, Japan, 8–10 September 2016. *Procedia Computer Science.* 2017;111:413–419.

43. Balo F, Şağbanşua L. Wind turbine selection for energy efficiency: Multi-criteria decision making, 23. In: *International Energy and Environment Fair and Conference (ICCI 2017),* İstanbul, 28–31, 3–5 May 2017.

44. Sua LS, Balo F. Techno-economic model for optimum design of wind energy facilities. In: *International Engineering and Technology Symposium, IETS 2018,* Batman, Turkey, 3–5 May 2018.

45. Department of Wind Energy TU. WAsP 10 help facility and on-line documentation. Denmark, 24 February 2012.

46. Frank HP, Rathmann O, Mortensen NG, Landberg L. The numerical wind atlas—The KAMM/WAsP method. Riso, Roskilde, Denmark, 2001.

47. Miljødata EO. Case studies calculating wind farm production—Main report. Energi-og Miljødata, Denmark, 2002.

48. Sajan AM, Mariappan VN. Wind resource land mapping using ArcGIS, WAsP and multi criteria decision analysis (MCDA). *Energy Procedia.* 2014;52:666–675. http://dx.doi.org/10.1016/j.egypro.2014.07.123.

49. http://www.tetasbilisim.com.tr/index.php/urunler/14-portfolio/13-yazilim -wasp.

50. Mortensen NG. Wind resource assessment using the WAsP software (DTU Wind Energy E-0135), Technical University of Denmark (DTU), DTU Wind Energy E, No, 0135, 2016.

51. https://www.sciencedirect.com/topics/engineering/weibull-probability -distribution.

52. Bataineh KM, Dalalah D. Assessment of wind energy potential for selected areas in Jordan. *Renewable Energy.* 2013;59:75–81.

53. Ohunakin OS, Akinnawonu OO. Assessment of wind energy potential and the economics of wind power generation in Jos, Plateau State, Nigeria. *Energy for Sustainable Development.* 2012;16:78–83.

54. https://tr.climate-data.org/asya/tuerkiye/diyarbak%C4%B1r/diyarbak%C4 %B1r-285/#climate-graph.

55. https://www.enerjiatlasi.com/ruzgar-enerjisi-haritasi/diyarbakir.

56. http://www.yegm.gov.tr/YEKrepa/DIYARBAKIR-REPA.pdf.

57. Wegley HL, Ramsdell JV, Orgill MM, Drake RL. A siting handbook for small wind energy conversion systems. Battelle Pacific Northwest Lab, 1980.

58. Sharma K, Ahmed MR. Wind energy resource assessment for the Fiji Islands: Kadavu Island and Suva Peninsula. *Renewable Energy.* 2016;89:168–180.

Chapter 5

Supercapacitor for Sustainable Energy Storage

Mingyuan Zhang, Zhaoru Shang, and Haozhe Yi

INTRODUCTION

Fully employing energy resources from nature has been important for human beings to develop new technologies and sustainable life. As society is growing at a high rate, the demand for electricity in all kinds of occasions is increasing dramatically. For the time being, people have been using lots of methods such as combustion of fossil fuels, gas and coal for thermal power (Bertine & Goldberg, 1971; Carlson & Adriano, 1993; Nikolich, 1983, 1984; Singer, 1991). Consequently, the negative effects associated with combustion bring severe damage to the environment (Dincer, 2000; Lélé, 1991). Therefore, people around the world are trying to develop more environment-friendly sustainable energy sources like hydropower, wind power, geothermal power, solar power, and other bioenergies (Baños et al., 2011; DiPippo, 2012; Dui et al., 2017; Kou et al., 2019a, 2020; Moran et al., 2018). The requirement of new-generation, highly efficient, and clean energy storage devices has become a priority to society. Rechargeable batteries, supercapacitors, solar cells, and other hydrogen storage devices have become a certain fields for scientists to study (Carlson & Wronski, 1976; Goodenough & Park, 2013; Mebratu, 1998; Murray et al., 2009; Züttel, 2003).

Diverse batteries have been invented to be adapted for various kinds of electricity storage and conversion environments. However, batteries are still facing some challenges, and different energy storage and conversion systems are bought out (Meissner & Richter, 2005). Supercapacitors are designed to be the energy storage devices that may replace and improve batteries in some circumstances (Gidwani et al., 2014). The unique advantages of supercapacitors include, but not limited to, lightweight, nontoxic components, high power density, long-lasting cycling stability, and very fast charging time which

99

enables supercapacitors to be an excellent candidate for next-generation vehicle power resources, portable energy storage systems, energy conversion systems, etc. (Alva et al., 2018; Brockway et al., 2019; Conte, 2010; Dincer & Rosen, 2002; Kou et al., 2019b, 2020; Olabi, 2017; Ribeiro et al., 2001).

Supercapacitor is an electricity storage system that stores electrical charges on the surface electrode-electrolyte interface (Lee & Goodenough, 1999; Wang et al., 2009; Zhang & Zhao, 2009; Zhao & Zheng, 2015; Zhong et al., 2019). The first patent on the supercapacitor system was given to H. I. Becker from General Electric in 1957, which employed the double-layer charge storage mechanism (Becker, 1957). However, the first supercapacitor system was cumbersome and inefficient, so it was impractical to use it in real applications. The prototype of today's supercapacitors was invented by Robert A. Rightmire from Standard Oil Company of Ohio (Rightmire, 1966). After that, the supercapacitor system attracted increasing attention from international research groups and commercial companies (Conte, 2010).

Nowadays, considering the growing demand for low-carbon economy and sustainability for the society, supercapacitors have proven to be a promising energy storage technology for sustainable energy resource (Gao et al., 2012; Zhao & Zheng, 2015; Zhu et al., 2020). The electrode material is the key factor to improve the power density, increase the cycling ability, and accelerate the charging rate for supercapacitors (Gao et al., 2012; Portet et al., 2005). Carbon is a lightweight, nontoxic, and easy-to-obtain material for industrial and energy applications that can be effectively utilized as the basic electrode material for supercapacitors (Obreja, 2008). Furthermore, some pseudocapacitive materials can induce redox reaction to the chemical process of supercapacitors and further improve the power density and charging efficiency (Chen et al., 2014; Hou et al., 2014; Lu et al., 2013). For pseudocapacitors, metal oxides are most widely used materials (Zhang & Chen, 2008). Besides this type of materials, nanomaterials with different morphologies and surface properties are also important research areas to improve supercapacitor performance (Chen et al., 2011; Saha et al., 2018; Yu et al., 2015).

In this chapter, we will primarily focus on their working mechanism, how carbon materials could benefit the usage of electrodes in supercapacitors, and the developments and cutting-edge research for those energy storage applications.

WORKING MECHANISM OF SUPERCAPACITORS AND THEIR MATERIALS

In the energy storage area, specific energy and specific power are two of the most important facts that people are concerned about. Specific energy describes the energy density, indicating how much energy can be stored per unit mass of active materials (Züttel, 2003). Sometimes the energy density stands for energy per unit volume. For example, batteries with higher specific energy usually promise a longer mileage range for electrical vehicles. Specific power is the output power per unit mass. It describes how fast energy can be released from the energy storage devices. If we use the same example, the specific power basically determines the acceleration of the electric vehicles. Fuel cells and lithium batteries tend to have good specific energy, but relatively poor specific power. The traditional capacitors have excellent specific power (fast charging-discharging rate), however, they have low specific energy. The balance of these two parameters was gained by the development of the supercapacitors, which are sometimes called ultracapacitors or electric double-layer capacitors (EDLC). This is the early beginning of the discovery of supercapacitors which filled gap between conventional electrolytic capacitors and early batteries with relatively high power density and energy density.

The idea of supercapacitors (also known as ultracapacitors) originated since the discovery of double-layer theory in surface and colloid science in early 1879 by Helmholtz (Helmholtz, 1879). The theory states that free charge in the ionic liquid phase can be captured by the opposite charge on the metallic surface with the contact of the ionic liquid. The double-layer consists of an inner layer of ions close to the charging electrode and an outer layer of ions attached to the inner layer. The two layers are oppositely charged and balanced with each other. The double-layer structure enables the high capacity of the energy storage of EDLC. The significant capacitance increase of the supercapacitors is also attributed to the large surface area of the electrode materials. One representative example is activated carbon (AC). The large surface area allows a high capacity of energy storage.

To further increase the capacitance, pseudocapacitor was developed. Unlike traditional capacitors and EDLC, redox reaction happens in the pseudocapacitor during charging-discharging process. With the redox reaction, the pseudocapacitor has higher specific energy, but lower specific power than EDLC (Lu et al., 2013). To improve the specific power, active materials in nanostructure have been studied as a possible solution (Lu et al., 2013). Nanoscale particles have a shorter diffusion distance compared with large particles. Short diffusion distance facilitates fast sorption and desorption of ions, which means fast charging and discharging rate.

Supercapacitors have three major kinds in general for their energy storage mechanism: electrical double-layer capacitors, pseudocapacitors, and hybrid capacitors. Depending on the electrodes and electrolyte, they are also classified as symmetric supercapacitors and asymmetric supercapacitors with aqueous electrolyte, organic electrolyte, and ionic liquid electrolyte. Here, we mainly discuss the theory of EDLCs which are considered the footstone for all kinds of supercapacitors since it's highly necessary to understand the working mechanism for the preparation of cell design.

The Discovery of Double-Layer Theory

The most used supercapacitor was EDLC. It's a very unique energy storage device that does not require any chemical reaction but only utilizes the physical adsorption of ions to store charge on the electrode-electrolyte surface. Due to this mechanism, the speed of energy storage and release is relatively high compare to the electrochemical energy storage devices.

At the very beginning, it was commonly known that the model to calculate double-layer capacitance followed the theory developed by Helmholtz (1879). He discovered that when charged electrodes are immersed into the electrolyte, the same amount of opposite charges distributes on the two sides of the surface between electrolyte and electrode. The rules can be summed as follow:

$$\sigma = \frac{\varepsilon\varepsilon_0}{d} V \tag{5.1}$$

$$C_d = \frac{\partial\sigma}{\partial V} = \frac{\varepsilon\varepsilon_0}{d} \tag{5.2}$$

where, σ is the charge density, V is the voltage difference between two layers of charge, ε is the dielectric constant of the electrolyte in between, ε_0 is the vacuum permittivity, d is the spacing between two layers of charge, and C_d is the calculated differential specific capacitance per unit area.

This theory states that the specific capacitance remains constant regardless of the kind of electrolyte and the applied voltage across electrodes which lacks an explanation for all situations, because in Helmholtz's model, he considered that the ions in the electrolyte directly contacting with the surface act like a double-plate capacitor without considering the ion concentration in rest of the electrolyte (Helmholtz, 1879).

After a short period of discovery, another carefully modified model was developed by Louis Georges Gouy in 1910 and David Leonard Chapman in 1913 independently to further elucidate the relationship between the differential capacitance with more dependence like the concentration of the electrolyte and the applied voltage (Chapman, 1913; Gouy, 1910). They stated that there is an ion concentration distribution along the direction perpendicular to the interface of the electrolyte-electrode. When the surface charges and ions contribute to most of the energy storage, the ions with relatively low concentration in other parts of the electrolyte still need to counterbalance the excess amount of charges built up near the surface of the electrode and the thickness of the distribution layer is finite and called diffuse layer. For example, lower concentration electrolyte needs a larger amount of solution to counterbalance the excess charge, so the diffuse layer in the lower concentration electrolyte would be thicker than in the higher concentration electrolyte (Bard & Faulkner, 2001b).

One could equally divide the electrolyte perpendicular to the surface of the electrode to parallel layers which fall under the condition of thermal equilibrium. Set the bulk solution concentration as a reference, the number of ions in a random layer follows the Boltzmann distribution (Bard & Faulkner, 2001b). The total charge per volume at any given position apart from the electrode could be the summation of the total electron number of different kinds of ions multiplied by their valance number.

The charge density and electrostatic potential at distance x follow the Poisson equation:

$$\rho(x) = -\varepsilon\varepsilon_0 \frac{d^2\phi}{dx^2} \tag{5.3}$$

After derivation, the potential profile is as followed (Bard & Faulkner, 2001b):

$$\frac{d\phi}{dx} = -\left(\frac{8kTn^0}{\varepsilon\varepsilon_0}\right)^{1/2} \sinh\left(\frac{ze\phi}{2kT}\right) \tag{5.4}$$

$$\phi = \phi_0 e^{-\kappa x} \text{ where } \kappa = \left(\frac{2n^0 z^2 e^2}{\varepsilon\varepsilon_0 kT}\right)^{1/2} \tag{5.5}$$

By applying Gaussian law and differentiating the equation above, we could get the specific differential capacitance as (approximation at room temperature) (Bard & Faulkner, 2001b):

$$C_d = \frac{d\sigma^M}{d\phi_0} = 228zC^{*1/2} \cosh\left(19.5z\phi_0\right) \tag{5.6}$$

where, σ^M is the excess charge density and C^* is the bulk electrolyte concentration, all in SI units.

With this correction, the equation predicts the differential capacitance more accurately. It occurs in a V-shape which relatively match and explain the experimental data. But there are still some drawbacks to this model. When the electrostatic potential increase to a certain large amount, the differential capacitance appeared to be infinite and failed to describe higher electrolyte concentration (Bard & Faulkner, 2001b).

The reason for this misinterpretation of the capacitance is mainly because the theory states that ions can appear at any point regardless of the ion radius. By considering this, the plane at the minimum distance away from the center of the ion to the electrode called the outer Helmholtz plane (OHP) is developed by Stern. This modification is still based on Gouy-Chapman theory and can be applied with given conditions. For example, in a very–low-concentration electrolyte, the OHP would have nearly no effect on the calculated capacitance because the diffuse layer would be large enough to overlook this thickness. For a relatively high concentration electrolyte, because the diffuse layer is compact, the thickness of this layer would not be negligible anymore.

For the distance away from the OHP, the Poisson-Boltzmann equation could still apply after simplification with the form of (Bard & Faulkner, 2001b):

$$\frac{\tanh\left(ze\phi / 4kT\right)}{\tanh\left(ze\phi_2 / 4kT\right)} = e^{-\kappa(x-x_2)} \tag{5.7}$$

where, ϕ_2 is the electrical potential at the position of OHP and κ is defined above.

For the distance within the OHP to the electrode surface, because there's no charge in between, the electrical potential profile would be linear. By applying Gauss law again, we could get the total differential capacitance (Bard & Faulkner, 2001b). The total capacitance C_d consists of two capacitances (The capacitance within the OHP, C_H, and the capacitance in the diffuse layer, C_D) and we could treat them as series connection. The equation after arrangement and simplification is followed (Bard & Faulkner, 2001b):

$$\frac{1}{C_d} = \frac{1}{C_H} + \frac{1}{C_D} = \frac{x_2}{\varepsilon\varepsilon_0} + \frac{1}{\left(2\varepsilon\varepsilon_0 z^2 e^2 n^0 / kT\right)^{1/2} \cosh\left(ze\phi_2 / 2kT\right)} \tag{5.8}$$

From the equation, we could see that C_H is independent of electrical potential. C_D is a hyperbola curve associated with electrical potential. For lower concentration, C_D is small enough that the shape of the C_d curve still remains a similar shape as C_D. For higher concentration, C_D would be much larger than C_H so the curve would become a flat shape which only shows the dependency of C_H.

With these modifications added to the previous equations, the new expression for the differential capacitance is the so-called Gouy-Chapman-Stern model. Although this model could explain the experimental data with better accuracy, there are still existing problems that need to be solved with more consideration. For example, the influence of ions to each other, the solvated ion radius, and other possible reactions that happen on the electrode.

With new kinds of electrode-electrolyte systems appearing, Kornyshev also proposed the theory in 2007 to describe the interfacial capacitance between the planer metal with ionic liquid interface (Kornyshev, 2007) and Goodwin added some modifications with accounting the short-range correlations afterward (Goodwin et al., 2017). Other models and methods that could also be used to describe the behaviors of double-layer capacitors afterward surpassed the limits of classic electrostatics such as using molecular dynamics, Monte Carlo methods, and density functional theory (Zhan et al., 2017).

Electrode Materials

Supercapacitors usually have three key components which are electrode, separator, and electrolyte. These three parts hold the structural integrity and also determine the performance of the supercapacitors. Considering the growing demand for energy efficiency and environmental sustainability, the advantage of supercapacitors stood out with having much faster-charging speed, longer cycling time, wider operating temperature, and utilizing more environment-friendly materials than rechargeable batteries which also share relatively same key components. Supercapacitors usually have higher power density and relative lower energy density than rechargeable batteries, but with proper cell design, the specific power and specific energy density can increase with a large magnitude, making them have wider applications (Pandolfo & Hollenkamp, 2006).

Low-carbon economy requires avoiding the production of greenhouse gas. Metallic materials such as copper, iron are no longer the first priority in the industry for the developments in technology (Raj et al., 2017). Carbon has become the newly developed key technology in all kinds of industrial applications. Especially in energy development, carbon materials play an important role in the energy storage field such as batteries and supercapacitors. Carbon is the most common electrode material used in supercapacitors, particularly in EDLCs which are also the major part for supercapacitors in the market now.

EDLCs utilize various kinds of carbonaceous materials as electrodes like carbon black, activated carbon, carbon fiber, carbon nanotubes (CNTs), Graphene, vitreous carbon (glassy carbon), and other carbon networks (An et al., 2001; Gamby et al., 2001; Kossyrev, 2012; Le et al., 2013; Yuan et al., 2013). The reason why carbonaceous materials are more favorable to electrodes is because of their porous structure and high specific surface area (SSA) (Saliger et al., 1998; Sircar et al., 1996). The porous structure usually enables an extremely large surface area that electrolyte could enter which provides the space for opposite charge ions to adsorb onto the electrode to store energy. We are going to mainly discuss two classic carbon electrodes for EDLCs that match the requirements above.

Carbon black is a very typical form of nanocarbon (Coville et al., 2011). It is usually made by incomplete combustion of gaseous or liquefied hydrocarbons or thermal decomposition under certain pressure and temperature (Llobet, 2020). Carbon black is a form of pure carbon usually comes with the shape of the colloid spherical particle. It has been widely used in all kinds of applications like pigment, nanofiller reinforcement materials in rubber products to further increase the mechanical strength because it can increase the bonding strength during the forming of compound (Farida et al., 2019). Other applications in electronics are also developed because of their good electrical conductivity as well as the large surface area which is up from 20 to 1500 m^2/g. Usually, smaller spherical carbon black has a darker color and higher viscosity and lower wettability which offers higher conductivity and greater absorption in UV radiation (Pantea et al., 2003; Probst & Grivei, 2002; Tchoudakov et al., 1996).

Electrically conductive carbon black usually comes with a particle size of 10–100 nm which is a perfect filler material for energy storage application. The spherical and fined particle could filtrate into the binder content and form a "bridge" between adjacent larger particles, which significantly improve the electrical conductivity by electrons jumping across the gap between closed-pack aggregates (Pandolfo & Hollenkamp, 2006). Carbon black nanoparticles tend to aggregate due to their high surface energy and the electrostatic charge they carried. With a certain amount of loading beyond a percolation threshold in the electrode, carbon black nanoparticles could lead to an abrupt change in electrical conductivity and reach a plateau (Pandolfo & Hollenkamp, 2006).

Some electrodes also utilize carbon black as the main composition during processing. Commercialized porous carbon electrodes use a specific type of carbon black (Black Pearl 2000, Vulcan XC72R) mixing with 10–15% amount of polymer binder (PTFE/PVDF) to achieve the structural integrity of the electrodes. The binder content needs to uniformly disperse into the carbon black powder with alcohol solution. After thoroughly mixing and evaporating the solution in the oven, the membrane-like nanocomposite carbon black electrode is formed. The specific capacity for those commercialized carbon black electrode from Material Methods (PACMM™210) could reach to 113

F/g in 1M H_2SO_4 electrolyte and other electrodes produced by high surface area could reach to 250 F/g (Beck et al., 2001). The pore size distribution of the carbon black/acetylene black polymer composite electrode ranges from less than 2 nm (micropores) to more than 50 nm (mesopores), and most of the pores are contributed by micropores which provide a sufficient surface area for the ions to pair.

The advantage of carbon black composite electrode is that they are generally easy to prepare so that the cost for the product is reduced. The available SSA is mostly contributed by the spherical and solid carbon black powder surface. In order to get higher surface area, the particle should be smaller and smaller, which give rise to a severe problem in aggregation and robustness of the structure of electrodes (Rennie et al., 2016). Also, the volumetric power density for carbon black is low compared to other carbonaceous materials because carbon black has an extremely low density which is not suitable for some designs of compact electric power sources that require high energy density as well as power density (Frackowiak, 2007).

Another classic carbon material for supercapacitors is activated carbon, which is the most common carbon material in electrode fabrication. It has a much higher SSA (>1500 m^2/g) than most of the solid spherical carbon black. Different from most of the carbon black nanoparticles, powdered activated carbon has huge number of micropores within each particle due to the activation method. The pore size distribution for activated carbon usually is micropores (<2 nm) and with a certain amount of mesopore (2–50 nm) and macropores (>50 nm). Micropores mainly contribute to most of SSA for ions adsorption while mesopores and macropores create the pathway and act like tunnels for ions in the electrolyte to have access to those micropores. It is noteworthy that for pores smaller than the solvated ion size, the ions which cannot enter the pores could lead to low specific capacitance (Largeot et al., 2008).

Activated carbon can be derived from various sources in many forms using different activation methods. The precursors of producing activated carbon could be bio-waste materials such as coconut shell, palm shell, biochar, charcoal, and other agricultural residues (Azargohar & Dalai, 2006; Daud & Ali, 2004; Hassan et al., 2020), the precursors are easy to obtain and some process could even be free of greenhouse gas emission.

The activation methods could be generally divided into physical activation and chemical activation. For physical activation, certain kinds of oxidizing gases like oxygen, carbon dioxide, and air are used (Jänes et al., 2007). The bio-waste carbon source is generally exposed to pyrolysis in an inert atmosphere at 400–900°C to burn off volatile species followed by the combustion of active gases to open up close pores to form interconnected microporous structures. By modulating the temperature, active gas agents could achieve control of pore size distribution. For the chemical activation process, heat treatment is needed at 450–900°C. The advantage of chemical activation is that it only needs one step to achieve even higher porous structure than the physical process. KOH is generally used as an oxidant agent for this process (Wang & Kaskel, 2012).

Activated carbon comes in many forms such as powdered, bead, fiber, and cloth (Hassan et al., 2020). For activated carbon powder, the electrode fabrication method remains the same as the carbon black powder mentioned above. Electrically conductive carbon black now is added as conducting and filler agents to fill the gap between activated carbon powder and binder content to enhance the conductivity as well as lower the impedance for the cell (Son et al., 2020). Activated carbon cloth (ACC) is generally used more in solid-state supercapacitors. Because of its higher strength and robustness, more applications are focus on the stretchable supercapacitors which will be discussed in the following sections.

STATE-OF-THE-ART DEVELOPMENTS OF SUPERCAPACITORS

Conducting-Polymer-Based Supercapacitors

Conducting polymers have been studied for many years as potential materials for the supercapacitor electrodes. The supercapacitors with electrodes made from the conducting polymers are referred to pseudocapacitors, since their capacitance is based on the redox reaction of the conducting polymers.

Advantages and Challenges of
Conducting-Polymer-Based Supercapacitors

The conductivity, as an inherent property of the conducting polymers, covers a moderately large range (Lota et al., 2004; Mastragostino et al., 2001). Chemical oxidation or electrochemical oxidation of the monomers create the connections on the polymer backbone (Snook et al., 2011). The charging and discharging are realized by doping and undoping processes. For conducting polymers, there are mainly two forms of reactions which are p-doping and n-doping. The p-dopable and n-dopable polymers are used in the electrodes to create several types of supercapacitor devices. The most studied configuration of the conducting-polymer-based supercapacitor device consists of a negative electrode made of n-dopable polymer and a positive electrode made of p-dopable polymer. Theoretically when electrodes at both sides are being doped, a high operating voltage can be reached, and high specific energy and power are possible. However, the low efficiency of n-doping process due to the inherent high impedances may result in a lower capacitance than the expectation. Therefore, carbon-based or lithium-based negative electrodes are paired with the conducting-polymer-based positive electrode to form hybrid supercapacitors (Snook et al., 2011).

One disadvantage of the conducting-polymer-based supercapacitors is the relatively poor cycling life. Compared with more than half million cycles that the carbon-based supercapacitors can reach, the cycling life of the conducting-polymer-based supercapacitors is only a few thousand cycles (Snook et al., 2011). The doping-undoping process causes the swelling and contraction of the polymers during the charging-discharging loops. Due to the repeated volume change, the physical structure of polymers is hard to keep after about a thousand cycles. Degradation happens thereafter and these keep getting worse until unacceptance. Short lifetime means a frequent change of the devices, which will eventually raise the cost and lower the interest of the product in the market. The good news is that the low cost and the relatively high energy density after charging of conducting polymers can compensate for the cost issue caused by the short lifetime. Besides, the conducting polymers are easy to process in terms of manufacture. The polymer processing technologies are very mature, which can further reduce the cost of the conducting-polymer-based supercapacitors at the industry level.

Most Commonly Studied Conductive Polymers

Table 5.1 lists the conducting polymers that are commonly used in pseudocapacitors. The dopant levels of these are generally equal or less than 0.5. The potentials are between 0.7 and 1.2 V. The theoretical specific capacitance ranged from 210 to 750 F/g. Poly(3,4-ethylenedioxythiophene) (PEDOT) has the highest molecular weight. All the conducting polymers listed in Table 5.1 will be discussed in this section.

Table 5.1 Properties of Most Commonly Investigated Electrically Conducting Polymers (Lota et al., 2004)

Electrically Conducting Polymer	Molecular Weight (g/mol)	Doping Level	Potential (V)	Theoretical Capacitance (F/g)
PAni	93	0.5	0.7	750
PPy	67	0.33	0.8	620
PTh	84	0.33	0.8	485
PEDOT	142	0.33	1.2	210

Polyaniline (PAni) is one of the most commonly studied polymers for material of supercapacitors (Hussain et al., 2006; Kulesza et al., 2006; Ryu et al., 2002b; Sivakkumar & Saraswathi, 2004; Talbi et al., 2003). The reason that it attracted much attention is because of its high electroactivity, good dopant level, relatively good capacitance, and high electrochemical stability (Talbi et al., 2003). The redox reactions for the conducting polymers are generally fast, which facilitates a good charging-discharging rate. The dopant level of PAni is 0.5, which means that each dopant is formed by two monomers of the polymer. Compared with the other commonly studied conducting polymers, the dopant level of PAni is relatively high. Its reported capacity varies in a wide range (Sivakkumar & Saraswathi, 2004). The highest reported value was 270 mAh/g, which is comparable with the capacity of lithium-ion batteries. Compared with the specific power and specific energy of about 3 kW/kg and 5 Wh/kg for the supercapacitors composed of carbon electrodes on both side, the conducting-polymer-based supercapacitor can promise a specific power of 2 kW/kg and a doubled specific energy value (more than 10 Wh/kg) (Talbi et al., 2003).

Polypyrrole is another material that has been extensively studied (Boyano et al., 2007a, 2007b; Fan & Maier, 2006; Hussain et al., 2005; Snook et al., 2004; Snook & Chen, 2008; Tripathi et al., 2006; Wang et al., 2006, 2007a, 2007b). Compared with the other commonly used conducting polymer materials, this material provides better electrochemical processing flexibility (Hughes et al., 2004). The drawback of polypyrrole is that this material is not n-dopable, so that it is only used in cathodes. Doping with single-charged anions is common for polypyrrole. When doped with multiple-charged anions, microporous structures were created by the physical cross-linking. The diffusivity and capacitance were improved by physical change, making polypyrrole a good candidate for supercapacitor materials (Suematsu et al., 2000).

Polythiophenes and their derivatives also have been studied as the electrode materials in supercapacitors (Mastragostino et al., 2002; Nohma et al., 1995; Peng et al., 1998; Rudge et al., 1994; Ryu et al., 2002a). The good stability of polythiophenes and most of their derivatives in ambient conditions of both air and moisture allow them to be used in many practical applications (Lota et al., 2004). One of the commonly investigated derivatives is PEDOT. According to Lota et al. (Lota et al., 2004), composites consisted of PEDOT and multi-walled CNTs have capacitance between 60 and 160 F/g with good cycling capability in all the electrolytes. Due to the high material density of the PEDOT, its volumetric energy is excellent. In fact, rather than improving the specific energy, composites materials usually better improve the cycling performance (mainly by enhancing the material stability) and the material conductivity (Snook et al., 2011).

Flexible Supercapacitors

Supercapacitors show great potential for sustainable energy storage applications due to their rapid charge-discharge rate, high energy density and power density, and excellent cycling ability (Iro et al., 2016). As technology is progressing, sustainable energy storage devices are facing new challenges to be utilized in specific situations. Flexible supercapacitors are designed to be flexible and lightweight, which meets the requirement to be a promising component in new generations of wearable, portable, or flexible electronic products (Vangari et al., 2013). In recent decades, flexible supercapacitors have aroused great attention in the electrochemistry field. With the growth of flexible supercapacitors, novel designs of touch sensors or display technologies can be achieved and flexible electronics can be realized (Dong et al., 2016). In order to get a flexible capacitor with high performance and environment-friendly design, special materials as well as new device configurations are supposed to be considered. Among them, developing flexible positive and negative electrodes, as well specific electrolytes have become the most crucial parts for flexible supercapacitors.

Flexible Electrodes

Currently, most of supercapacitors in the industry are made of rigid electrodes (Wang et al., 2012). Choosing suitable flexible electrodes has become most challenging part in the design of supercapacitors. For flexible supercapacitors, not only the electrochemistry behavior but mechanical performance of electrodes is also considered, since the active material should not break down during bending or folding. At this stage, most of the flexible electrodes are made up with electrochemistry active thin films and soft substrates.

For supercapacitors with various mechanisms, electrode materials are different. In general, there are two types of electrochemical capacitors used for sustainable energy storage: EDLCs and pseudocapacitors (González et al., 2016). Metal oxides and conducting polymers are wildly used electrode materials for pseudocapacitors (Zhi et al., 2013). While for EDLCs, carbon materials are the dominating electrode materials which are promising candidates for the flexible supercapacitors since carbon materials with various morphologies and dimensions have been extensively studied and confirmed to have extraordinary performances for EDLCs (Gwon et al., 2011). Up until now, various carbon nanomaterials, such as fullerene (zero dimension), CNTs (one dimension), and graphene (two dimension), and diverse fabrication methods have been tested for flexible supercapacitors electrodes (Shi et al., 2013).

In assistance of hydrogen bonds or van der Waals forces, carbon nanomaterials can aggregate into carbon networks which can directly serve as flexible electrodes in supercapacitors. For instance, through waving, printing technology, chemical vapor deposition or dipping-drying process, CNTs or graphene can be turned into carbon fabrics, clothes, films, papers, or textiles (Hu et al., 2010; Masarapu et al., 2012; Meng et al., 2010; Nyholm et al., 2011; Yu et al., 2009). These soft, foldable, and stretchable single-carbon materials exhibit good strength, stiffness and flexibility, and can be directly applied in flexible capacitors.

Despite the fabrication and application of single-carbon nanomaterials have been well studied and improved in recent years, single-carbon materials are not good enough to be involved in reversible redox reactions so that they can only serve as EDLC electrodes. Therefore, in order to improve the specific capacitance, pseudocapacitive materials are coated on flexible substrate like single-carbon architectures so that flexible pseudocapacitors electrodes are realized. Pseudocapacitive materials like metal oxides, metal nitrides, or metal sulfides are designed to be coated on flexible materials like carbon fabrics or conducting polymers to realize this function (Wang et al., 2012).

Electrolytes in Flexible Supercapacitors

Besides flexible electrodes, electrolyte is another important component in flexible supercapacitors. Generally, there are two types of electrolytes utilized in flexible supercapacitors: liquid electrolytes and solid-state electrolytes (Zhong et al., 2015). Liquid electrolytes for flexible supercapacitors are pretty much the same as conventional supercapacitors.

In order to keep the structural stability and electrochemistry performance of flexible supercapacitors after deformation, solid-state electrolytes are more desirable since they not only transfer ions but also provide structural support. Solid-state electrolytes can be classified into three categories, which are aqueous polymeric gels, nonaqueous polymeric gels, and inorganic solid materials (Yang & Mai, 2014). Most of the solid-state electrolytes are aqueous or nonaqueous polymeric gels produced by sol-gel method which refers to a mixture of aqueous or organic solutes, solvents, and gel agents. Inorganic solid electrolyte materials receive less attention compared with aforementioned two types of materials since most of the inorganic materials are not bendable. However, inorganic materials are thermally stable, which enables them for specific utilizations. For example, Francisco et al. presented a new kind of flexible and Li-ion-conductive material, $Li_2S-P_2S_5$ glass-ceramic, which can be used in all solid-state flexible supercapacitors (Francisco et al., 2012).

With the progressing of flexible supercapacitors, new generations of electronics can expect extraordinary flexible mechanical performances. Flexible supercapacitors can be a reliable power source due to their outstanding charge-discharge rate, high power density, as well as deformable structures.

Flexible electrodes for supercapacitors have been extensively studied and multifarious materials have been approved. However, most of the electrode materials utilized in supercapacitors at this stage are carbon-based materials which are not pseudocapacitive. How to make pseudocapacitive materials flexible without damaging the electrochemistry performance remains a key issue in flexible supercapacitor development. Appropriate solid-state electrolytes and desirable supercapacitor configurations are also crucial fields in the development of solid-state flexible supercapacitors.

CONCLUSION AND FUTURE OUTLOOK

Supercapacitor is an emerging energy storage device which fully utilizes the outstanding benefits of carbon materials. The ultra-high charging speed and power density due to fast adsorption and release of ions makes it a future-promising battery replacement. With the advantages of long-lasting cycle stability, environmental friendliness, and vast application prospects, it is highly needed to spend more time and money to further study this field of research by trying to overcome the drawback of relatively low energy density before it could fully replace batteries in electric vehicles, telecommunications, and aerospace engineering with lower cost. Moreover, EDLCs still account for most of the market share for the whole supercapacitors market in 2019 and it's widely used in the field that requires steady energy supply over a short period of time (*Supercapacitor Market with COVID-19 Impact Analysis by Type, Electrode Material, Application, Region—Global Forecast to 2025*, 2020). The environmental demand for electric vehicles over traditional vehicles has highly increased the demand for supercapacitors-based vehicles over the battery-powered vehicles, as well as other applications like self-powered systems or maybe laptops and smartphones since they also require high charging speed. The total supercapacitors market for 2019 was valued at 426 million USD and it is predicted to reach 750 million USD in 2025 (*Supercapacitor Market with COVID-19 Impact Analysis by Type, Electrode Material, Application, Region—Global Forecast to 2025*, 2020), so it's highly necessary to further carry on the study of supercapacitors. Apart from EDLCs, pseudocapacitors and hybrid capacitors are also worthy choices to further boost the energy density by involving Faradaic reactions. With the combination of conducting polymer, functionalized ACC with transition metal oxides, the capacitance and mechanical strength could be significantly improved. The next action to take is to enhance the stability of conducting polymer composites as well as mitigate the usage of materials with high burden to environment as well as developing clean energy source to accommodate the imperative demand of sustainable everyday life.

REFERENCES

Alva, G., Lin, Y., & Fang, G. (2018). An overview of thermal energy storage systems. *Energy, 144,* 341–378.

An, K. H., Kim, W. S., Park, Y. S., Choi, Y. C., Lee, S. M., Chung, D. C., Bae, D. J., Lim, S. C., & Lee, Y. H. (2001). Supercapacitors using single-walled carbon nanotube electrodes. *Advanced Materials, 13*(7), 497–500.

Azargohar, R., & Dalai, A. K. (2006). Biochar as a precursor of activated carbon. *Applied Biochemistry and Biotechnology, 131*(1), 762–773. https://doi.org/10.1385/ABAB:131:1:762

Baños, R., Manzano-Agugliaro, F., Montoya, F. G., Gil, C., Alcayde, A., & Gómez, J. (2011). Optimization methods applied to renewable and sustainable energy: A review. *Renewable and Sustainable Energy Reviews, 15*(4), 1753–1766. https://doi.org/10.1016/j.rser.2010.12.008

Bard, A. J., & Faulkner, L. R. (2001a). *Fundamentals and applications: Electrochemical methods.* Wiley.

Bard, A. J., & Faulkner, L. R. (2001b). Fundamentals and applications. *Electrochemical Methods, 2*(482), 580–632.

Beck, F., Dolata, M., Grivei, E., & Probst, N. (2001). Electrochemical supercapacitors based on industrial carbon blacks in aqueous H_2SO_4. *Journal of Applied Electrochemistry, 31*(8), 845–853. https://doi.org/10.1023/A:1017529920916

Becker, H. I. (1957). *Low voltage electrolytic capacitor.* Google Patents.

Bertine, K. K., & Goldberg, E. D. (1971). Fossil fuel combustion and the major sedimentary cycle. *Science, 173*(3993), 233–235.

Boyano, I., Bengoechea, M., de Meatza, I., Miguel, O., Cantero, I., Ochoteco, E., Grande, H., Lira-Cantú, M., & Gomez-Romero, P. (2007a). Influence of acids in the Ppy/V_2O_5 hybrid synthesis and performance as a cathode material. *Journal of Power Sources, 174*(2), 1206–1211.

Boyano, I., Bengoechea, M., de Meatza, I., Miguel, O., Cantero, I., Ochoteco, E., Rodriguez, J., Lira-Cantú, M., & Gomez-Romero, P. (2007b). Improvement in the Ppy/V_2O_5 hybrid as a cathode material for Li ion batteries using PSA as an organic additive. *Journal of Power Sources, 166*(2), 471–477.

Brockway, P. E., Owen, A., Brand-Correa, L. I., & Hardt, L. (2019). Estimation of global final-stage energy-return-on-investment for fossil fuels with comparison to renewable energy sources. *Nature Energy, 4*(7), 612–621.

Carlson, C. L., & Adriano, D. C. (1993). Environmental impacts of coal combustion residues. *Journal of Environmental Quality, 22*(2), 227–247.

Carlson, D. E., & Wronski, C. R. (1976). Amorphous silicon solar cell. *Applied Physics Letters, 28*(11), 671–673.

Chapman, D. L. (1913). LI. A contribution to the theory of electrocapillarity. *The London, Edinburgh, and Dublin Philosophical Magazine and Journal of Science, 25*(148), 475–481.

Chen, S. M., Ramachandran, R., Mani, V., & Saraswathi, R. (2014). Recent advancements in electrode materials for the high-performance electrochemical supercapacitors: A review. *International Journal of Electrochemical Science, 9*(8), 4072–4085.

Chen, W., Rakhi, R. B., Hu, L., Xie, X., Cui, Y., & Alshareef, H. N. (2011). High-performance nanostructured supercapacitors on a sponge. *Nano Letters, 11*(12), 5165–5172.

Conte, M. (2010). Supercapacitors technical requirements fornew applications. *Fuel Cells, 10*(5), 806–818. https://doi.org/10.1002/fuce.201000087

Coville, N. J., Mhlanga, S. D., Nxumalo, E. N., & Shaikjee, A. (2011). A review of shaped carbon nanomaterials. *South African Journal of Science, 107*(3–4), 1–15.

Daud, W. M. A. W., & Ali, W. S. W. (2004). Comparison on pore development of activated carbon produced from palm shell and coconut shell. *Bioresource Technology, 93*(1), 63–69. https://doi.org/10.1016/j.biortech.2003.09.015

Dincer, I. (2000). Renewable energy and sustainable development: A crucial review. *Renewable & Sustainable Energy Reviews, 4*(2), 157–175. https://doi.org/10.1016/S1364-0321(99)00011-8

Dincer, I., & Rosen, M. (2002). *Thermal energy storage: Systems and applications.* John Wiley & Sons.

DiPippo, R. (2012). *Geothermal power plants: Principles, applications, case studies and environmental impact.* Butterworth-Heinemann.

Dong, L., Xu, C., Li, Y., Huang, Z. H., Kang, F., Yang, Q. H., & Zhao, X. (2016). Flexible electrodes and supercapacitors for wearable energy storage: A review by category. *Journal of Materials Chemistry A, 4*(13), 4659–4685. https://doi.org/10.1039/c5ta10582j

Dui, X., Zhu, G., & Yao, L. (2017). Two-stage optimization of battery energy storage capacity to decrease wind power curtailment in grid-connected wind farms. *IEEE Transactions on Power Systems, 33*(3), 3296–3305.

Fan, L. Z., & Maier, J. (2006). High-performance polypyrrole electrode materials for redox supercapacitors. *Electrochemistry Communications, 8*(6), 937–940.

Farida, E., Bukit, N., Ginting, E. M., & Bukit, B. F. (2019). The effect of carbon black composition in natural rubber compound. *Case Studies in Thermal Engineering, 16*, 100566. https://doi.org/10.1016/j.csite.2019.100566

Frackowiak, E. (2007). Carbon materials for supercapacitor application. *Physical Chemistry Chemical Physics, 9*(15), 1774–1785.

Francisco, B. E., Jones, C. M., Lee, S. H., & Stoldt, C. R. (2012). Nanostructured all-solid-state supercapacitor based on Li 2S-P 2S 5 glass-ceramic electrolyte. *Applied Physics Letters, 100*(10). https://doi.org/10.1063/1.3693521

Gamby, J., Taberna, P. L., Simon, P., Fauvarque, J. F., & Chesneau, M. (2001). Studies and characterisations of various activated carbons used for carbon/carbon supercapacitors. *Journal of Power Sources, 101*(1), 109–116.

Gao, Q., Demarconnay, L., Raymundo-Piñero, E., & Béguin, F. (2012). Exploring the large voltage range of carbon/carbon supercapacitors in aqueous lithium sulfate electrolyte. *Energy & Environmental Science, 5*(11), 9611–9617.

Gidwani, M., Bhagwani, A., & Rohra, N. (2014). Supercapacitors: The near future of batteries. *International Journal of Engineering Inventions (IJEI), 4*(5), 22–27.

González, A., Goikolea, E., Barrena, J. A., & Mysyk, R. (2016). Review on supercapacitors: Technologies and materials. *Renewable and Sustainable Energy Reviews, 58*, 1189–1206. https://doi.org/10.1016/j.rser.2015.12.249

Goodenough, J. B., & Park, K.-S. (2013). The Li-ion rechargeable battery: A perspective. *Journal of the American Chemical Society*, *135*(4), 1167–1176.

Goodwin, Z. A. H., Feng, G., & Kornyshev, A. A. (2017). Mean-field theory of electrical double layer in ionic liquids with account of short-range correlations. *Electrochimica Acta*, *225*, 190–197.

Gouy, M. (1910). *Sur la constitution de la charge électrique à la surface d'un électrolyte.* J. Phys. Theor. Appl., 9(1): 457–468

Gwon, H., Kim, H. S., Lee, K. U., Seo, D. H., Park, Y. C., Lee, Y. S., Ahn, B. T., & Kang, K. (2011). Flexible energy storage devices based on graphene paper. *Energy and Environmental Science*, *4*(4), 1277–1283. https://doi.org/10.1039/c0ee00640h

Hassan, M. F., Sabri, M. A., Fazal, H., Hafeez, A., Shezad, N., & Hussain, M. (2020). Recent trends in activated carbon fibers production from various precursors and applications—A comparative review. *Journal of Analytical and Applied Pyrolysis*, *145*, 104715. https://doi.org/10.1016/j.jaap.2019.104715

Helmholtz, H. (1879). Studien über electrische Grenzschichten. *Annalen Der Physik*, *243*(7), 337–382. https://doi.org/10.1002/andp.18792430702

Hou, Y., Chen, L., Liu, P., Kang, J., Fujita, T., & Chen, M. (2014). Nanoporous metal based flexible asymmetric pseudocapacitors. *Journal of Materials Chemistry A*, *2*(28), 10910–10916.

Hu, L., Pasta, M., La Mantia, F., Cui, L., Jeong, S., Deshazer, H. D., Choi, J. W., Han, S. M., & Cui, Y. (2010). Stretchable, porous, and conductive energy textiles. *Nano Letters*, *10*(2), 708–714. https://doi.org/10.1021/nl903949m

Hughes, M., Chen, G. Z., Shaffer, M. S. P., Fray, D. J., & Windle, A. H. (2004). Controlling the nanostructure of electrochemically grown nanoporous composites of carbon nanotubes and conducting polymers. *Composites Science and Technology*, *64*(15), 2325–2331.

Hussain, A. M. P., Kumar, A., Singh, F., & Avasthi, D. K. (2006). Effects of 160 MeV Ni[12+] ion irradiation on HCl doped polyaniline electrode. *Journal of Physics D: Applied Physics*, *39*(4), 750.

Hussain, A. M. P., Saikia, D., Singh, F., Avasthi, D. K., & Kumar, A. (2005). Effects of 160 MeV Ni[12+] ion irradiation on polypyrrole conducting polymer electrode materials for all polymer redox supercapacitor. *Nuclear Instruments and Methods in Physics Research Section B: Beam Interactions with Materials and Atoms*, *240*(4), 834–841.

Iro, Z. S., Subramani, C., & Dash, S. S. (2016). A brief review on electrode materials for supercapacitor. *International Journal of Electrochemical Science*, *11*(12), 10628–10643. https://doi.org/10.20964/2016.12.50

Jänes, A., Kurig, H., & Lust, E. (2007). Characterisation of activated nanoporous carbon for supercapacitor electrode materials. *Carbon*, *45*, 1226–1233. https://doi.org/10.1016/j.carbon.2007.01.024

Kornyshev, A. A. (2007). *Double-layer in ionic liquids: Paradigm change?* ACS Publications.

Kossyrev, P. (2012). Carbon black supercapacitors employing thin electrodes. *Journal of Power Sources*, *201*, 347–352.

Kou, R., Zhong, Y., Kim, J., Wang, Q., Wang, M., Chen, R., & Qiao, Y. (2019a). Elevating low-emissivity film for lower thermal transmittance. *Energy and Buildings*, *193*, 69–77.

Kou, R., Zhong, Y., & Qiao, Y. (2019b). Effects of anion size on flow electrification of polycarbonate and polyethylene terephthalate. *Applied Physics Letters, 115*(7), 73704.

Kou, R., Zhong, Y., & Qiao, Y. (2020). Flow electrification of corona-charged polyethylene terephthalate film. *ArXiv Preprint ArXiv:2005.14385.*

Kulesza, P. J., Skunik, M., Baranowska, B., Miecznikowski, K., Chojak, M., Karnicka, K., Frackowiak, E., Béguin, F., Kuhn, A., & Delville, M.-H. (2006). Fabrication of network films of conducting polymer-linked polyoxometallate-stabilized carbon nanostructures. *Electrochimica Acta, 51*(11), 2373–2379.

Largeot, C., Portet, C., Chmiola, J., Taberna, P.-L., Gogotsi, Y., & Simon, P. (2008). Relation between the ion size and pore size for an electric double-layer capacitor. *Journal of the American Chemical Society, 130*(9), 2730–2731. https://doi.org/10 .1021/ja7106178

Le, V. T., Kim, H., Ghosh, A., Kim, J., Chang, J., Vu, Q. A., Pham, D. T., Lee, J.-H., Kim, S.-W., & Lee, Y. H. (2013). Coaxial fiber supercapacitor using all-carbon material electrodes. *ACS Nano, 7*(7), 5940–5947.

Lee, H. Y., & Goodenough, J. B. (1999). Supercapacitor behavior with KCl electrolyte. *Journal of Solid State Chemistry, 144*(1), 220–223.

Lélé, S. M. (1991). Sustainable development: A critical review. *World Development, 19*(6), 607–621. https://doi.org/10.1016/0305-750X(91)90197-P

Llobet, E. (2020). Chapter 4: Carbon nanomaterials. In E. B. T.-A. N. for I. G. M. Llobet (Ed.), *Micro and nano technologies* (pp. 55–84). Elsevier. https://doi.org/10 .1016/B978-0-12-814827-3.00004-9

Lota, K., Khomenko, V., & Frackowiak, E. (2004). Capacitance properties of poly (3,4-ethylenedioxythiophene)/carbon nanotubes composites. *Journal of Physics and Chemistry of Solids, 65*(2–3), 295–301.

Lu, Q., Chen, J. G., & Xiao, J. Q. (2013). Nanostructured electrodes for high-performance pseudocapacitors. *Angewandte Chemie International Edition, 52*(7), 1882–1889.

Masarapu, C., Wang, L. P., Li, X., & Wei, B. (2012). Tailoring electrode/electrolyte interfacial properties in flexible supercapacitors by applying pressure. *Advanced Energy Materials, 2*(5), 546–552. https://doi.org/10.1002/aenm.201100529

Mastragostino, M., Arbizzani, C., & Soavi, F. (2001). Polymer-based supercapacitors. *Journal of Power Sources, 97*, 812–815.

Mastragostino, M., Arbizzani, C., & Soavi, F. (2002). Conducting polymers as electrode materials in supercapacitors. *Solid State Ionics, 148*(3–4), 493–498.

Mebratu, D. (1998). Sustainability and sustainable development: Historical and conceptual review. *Environmental Impact Assessment Review, 18*(6), 493–520. https:/ /doi.org/10.1016/S0195-9255(98)00019-5

Meissner, E., & Richter, G. (2005). The challenge to the automotive battery industry: The battery has to become an increasingly integrated component within the vehicle electric power system. *Journal of Power Sources, 144*(2), 438–460.

Meng, C., Liu, C., Chen, L., Hu, C., & Fan, S. (2010). Highly flexible and all-solid-state paperlike polymer supercapacitors. *Nano Letters, 10*(10), 4025–4031. https:// doi.org/10.1021/nl1019672

Moran, E. F., Lopez, M. C., Moore, N., Müller, N., & Hyndman, D. W. (2018). Sustainable hydropower in the 21st century. *Proceedings of the National Academy of Sciences, 115*(47), 11891–11898.

Murray, L. J., Dincă, M., & Long, J. R. (2009). Hydrogen storage in metal–organic frameworks. *Chemical Society Reviews, 38*(5), 1294–1314.

Nikolich, M. (1983). *Combustion gas powered fastener driving tool.* Google Patents.

Nikolich, M. (1984). *Combustion gas-powered fastener driving tool.* Google Patents.

Nohma, T., Kurokawa, H., Uehara, M., Takahashi, M., Nishio, K., & Saito, T. (1995). Electrochemical characteristics of $LiNiO_2$ and $LiCoO_2$ as a positive material for lithium secondary batteries. *Journal of Power Sources, 54*(2), 522–524.

Nyholm, L., Nyström, G., Mihranyan, A., & Strømme, M. (2011). Toward flexible polymer and paper-based energy storage devices. *Advanced Materials, 23*(33), 3751–3769. https://doi.org/10.1002/adma.201004134

Obreja, V. V. N. (2008). On the performance of supercapacitors with electrodes based on carbon nanotubes and carbon activated material—A review. *Physica E: Low-Dimensional Systems and Nanostructures, 40*(7), 2596–2605. https://doi.org/10.1016/j.physe.2007.09.044

Olabi, A. G. (2017). *Renewable energy and energy storage systems.* Elsevier.

Pandolfo, A. G., & Hollenkamp, A. F. (2006). Carbon properties and their role in supercapacitors. *Journal of Power Sources, 157*(1), 11–27.

Pantea, D., Darmstadt, H., Kaliaguine, S., & Roy, C. (2003). Electrical conductivity of conductive carbon blacks: Influence of surface chemistry and topology. *Applied Surface Science, 217*(1), 181–193. https://doi.org/10.1016/S0169-4332(03)00550-6

Peng, Z. S., Wan, C. R., & Jiang, C. Y. (1998). Synthesis by sol–gel process and characterization of $LiCoO_2$ cathode materials. *Journal of Power Sources, 72*(2), 215–220.

Portet, C., Taberna, P.-L., Simon, P., Flahaut, E., & Laberty-Robert, C. (2005). High power density electrodes for carbon supercapacitor applications. *Electrochimica Acta, 50*(20), 4174–4181.

Probst, N., & Grivei, E. (2002). Structure and electrical properties of carbon black. *Carbon, 40*(2), 201–205. https://doi.org/10.1016/S0008-6223(01)00174-9

Raj, B., Van de Voorde, M., & Mahajan, Y. (2017). *Nanotechnology for energy sustainability, 3 volume set.* John Wiley & Sons.

Rennie, A. J. R., Martins, V. L., Smith, R. M., & Hall, P. J. (2016). Influence of particle size distribution on the performance of ionic liquid-based electrochemical double layer capacitors. *Scientific Reports, 6*, 22062.

Ribeiro, P. F., Johnson, B. K., Crow, M. L., Arsoy, A., & Liu, Y. (2001). Energy storage systems for advanced power applications. *Proceedings of the IEEE, 89*(12), 1744–1756.

Rightmire, R. A. (1966). *Electrical energy storage apparatus.* Google Patents.

Rudge, A., Raistrick, I., Gottesfeld, S., & Ferraris, J. P. (1994). A study of the electrochemical properties of conducting polymers for application in electrochemical capacitors. *Electrochimica Acta, 39*(2), 273–287.

Ryu, K. S., Kim, K. M., Park, N.-G., Park, Y. J., & Chang, S. H. (2002a). Symmetric redox supercapacitor with conducting polyaniline electrodes. *Journal of Power Sources, 103*(2), 305–309.

Ryu, K. S., Kim, K. M., Park, Y. J., Park, N.-G., Kang, M. G., & Chang, S. H. (2002b). Redox supercapacitor using polyaniline doped with Li salt as electrode. *Solid State Ionics*, *152*, 861–866.

Saha, S., Samanta, P., Murmu, N. C., & Kuila, T. (2018). A review on the heterostructure nanomaterials for supercapacitor application. *Journal of Energy Storage*, *17*, 181–202. https://doi.org/10.1016/j.est.2018.03.006

Saliger, R., Fischer, U., Herta, C., & Fricke, J. (1998). High surface area carbon aerogels for supercapacitors. *Journal of Non-Crystalline Solids*, *225*, 81–85.

Shi, S., Xu, C., Yang, C., Li, J., Du, H., Li, B., & Kang, F. (2013). Flexible supercapacitors. *Particuology*, *11*(4), 371–377. https://doi.org/10.1016/j.partic.2012.12.004

Simon, P., Gogotsi, Y., & Dunn, B. (2014). Where do batteries end and supercapacitors begin? *Science*, *343*(6176), 1210–1211.

Singer, J. G. (1991). *Combustion fossil power*. Combustion Engineering Inc.

Sircar, S., Golden, T. C., & Rao, M. B. (1996). Activated carbon for gas separation and storage. *Carbon*, *34*(1), 1–12.

Sivakkumar, S. R., & Saraswathi, R. (2004). Performance evaluation of poly (N-methylaniline) and polyisothianaphthene in charge-storage devices. *Journal of Power Sources*, *137*(2), 322–328.

Snook, G. A., & Chen, G. Z. (2008). The measurement of specific capacitances of conducting polymers using the quartz crystal microbalance. *Journal of Electroanalytical Chemistry*, *612*(1), 140–146.

Snook, G. A., Chen, G. Z., Fray, D. J., Hughes, M., & Shaffer, M. (2004). Studies of deposition of and charge storage in polypyrrole–chloride and polypyrrole–carbon nanotube composites with an electrochemical quartz crystal microbalance. *Journal of Electroanalytical Chemistry*, *568*, 135 142.

Snook, G. A., Kao, P., & Best, A. S. (2011). Conducting-polymer-based supercapacitor devices and electrodes. *Journal of Power Sources*, *196*(1), 1–12.

Son, I.-S., Oh, Y., Yi, S.-H., Im, W. B., & Chun, S.-E. (2020). Facile fabrication of mesoporous carbon from mixed polymer precursor of PVDF and PTFE for high-power supercapacitors. *Carbon*, *159*, 283–291. https://doi.org/10.1016/j.carbon.2019.12.049

Suematsu, S., Oura, Y., Tsujimoto, H., Kanno, H., & Naoi, K. (2000). Conducting polymer films of cross-linked structure and their QCM analysis. *Electrochimica Acta*, *45*(22–23), 3813–3821.

Supercapacitor Market. (2020). *Supercapacitor market with COVID-19 impact analysis by type, electrode material, application, region—Global forecast to 2025*. https://www.reportlinker.com/p04411873/Supercapacitor-Market-by-Type-Application-Vertical-and-Geography-Global-Forecast-to.html?utm_source=GNW

Talbi, H., Just, P.-E., & Dao, L. H. (2003). Electropolymerization of aniline on carbonized polyacrylonitrile aerogel electrodes: Applications for supercapacitors. *Journal of Applied Electrochemistry*, *33*(6), 465–473.

Tchoudakov, R., Breuer, O., Narkis, M., & Siegmann, A. (1996). Conductive polymer blends with low carbon black loading: Polypropylene/polyamide. *Polymer Engineering & Science*, *36*(10), 1336–1346. https://doi.org/10.1002/pen.10528

Tripathi, S. K., Kumar, A., & Hashmi, S. A. (2006). Electrochemical redox supercapacitors using PVdF-HFP based gel electrolytes and polypyrrole as conducting polymer electrode. *Solid State Ionics, 177*(33–34), 2979–2985.

Vangari, M., Pryor, T., & Jiang, L. (2013). Supercapacitors: Review of materials and fabrication methods. *Journal of Energy Engineering, 139*(2), 72–79. https://doi.org/10.1061/(ASCE)EY.1943-7897.0000102

Wang, G., Zhang, L., & Zhang, J. (2012). A review of electrode materials for electrochemical supercapacitors. *Chemical Society Reviews, 41*(2), 797–828. https://doi.org/10.1039/c1cs15060j

Wang, J., & Kaskel, S. (2012). KOH activation of carbon-based materials for energy storage. *Journal of Materials Chemistry, 22*(45), 23710–23725.

Wang, J, Wang, C. Y., Too, C. O., & Wallace, G. G. (2006). Highly-flexible fibre battery incorporating polypyrrole cathode and carbon nanotubes anode. *Journal of Power Sources, 161*(2), 1458–1462.

Wang, J., Xu, Y., Chen, X., & Du, X. (2007a). Electrochemical supercapacitor electrode material based on poly (3,4-ethylenedioxythiophene)/polypyrrole composite. *Journal of Power Sources, 163*(2), 1120–1125.

Wang, J., Xu, Y., Chen, X., & Sun, X. (2007b). Capacitance properties of single wall carbon nanotube/polypyrrole composite films. *Composites Science and Technology, 67*(14), 2981–2985.

Wang, X., Zhang, H., Zhang, J., Xu, H., Zhu, X., Chen, J., & Yi, B. (2006). A bifunctional micro-porous layer with composite carbon black for PEM fuel cells. *Journal of Power Sources, 162*(1), 474–479.

Wang, Y., Shi, Z., Huang, Y., Ma, Y., Wang, C., Chen, M., & Chen, Y. (2009). Supercapacitor devices based on graphene materials. *The Journal of Physical Chemistry C, 113*(30), 13103–13107.

Yang, P., & Mai, W. (2014). Flexible solid-state electrochemical supercapacitors. *Nano Energy, 8*, 274–290. https://doi.org/10.1016/j.nanoen.2014.05.022

Yu, C., Masarapu, C., Rong, J., Wei, B. Q. M., & Jiang, H. (2009). Stretchable supercapacitors based on buckled single-walled carbon nanotube macrofilms. *Advanced Materials, 21*(47), 4793–4797. https://doi.org/10.1002/adma.200901775

Yu, Z., Tetard, L., Zhai, L., & Thomas, J. (2015). Supercapacitor electrode materials: Nanostructures from 0 to 3 dimensions. *Energy & Environmental Science, 8*(3), 702–730.

Yuan, B., Xu, C., Deng, D., Xing, Y., Liu, L., Pang, H., & Zhang, D. (2013). Graphene oxide/nickel oxide modified glassy carbon electrode for supercapacitor and nonenzymatic glucose sensor. *Electrochimica Acta, 88*, 708–712.

Zhan, C., Lian, C., Zhang, Y., Thompson, M. W., Xie, Y., Wu, J., Kent, P. R. C., Cummings, P. T., Jiang, D., & Wesolowski, D. J. (2017). Computational insights into materials and interfaces for capacitive energy storage. *Advanced Science, 4*(7), 1700059.

Zhang, L. L., & Zhao, X. S. (2009). Carbon-based materials as supercapacitor electrodes. *Chemical Society Reviews, 38*(9), 2520–2531.

Zhang, S. W., & Chen, G. Z. (2008). Manganese oxide based materials for supercapacitors. *Energy Materials: Materials Science and Engineering for Energy Systems, 3*(3), 186–200. https://doi.org/10.1179/174892409X427940

Zhao, C., & Zheng, W. (2015). A review for aqueous electrochemical supercapacitors. *Frontiers in Energy Research, 3*(May), 1–11. https://doi.org/10.3389/fenrg.2015.00023

Zhi, M., Xiang, C., Li, J., Li, M., & Wu, N. (2013). Nanostructured carbon-metal oxide composite electrodes for supercapacitors: A review. *Nanoscale, 5*(1), 72–88. https://doi.org/10.1039/c2nr32040a

Zhong, C., Deng, Y., Hu, W., Qiao, J., Zhang, L., & Zhang, J. (2015). A review of electrolyte materials and compositions for electrochemical supercapacitors. *Chemical Society Reviews, 44*(21), 7484–7539. https://doi.org/10.1039/c5cs00303b

Zhong, Y., Kou, R., Wang, M., & Qiao, Y. (2019). Electrification mechanism of corona charged organic electrets. *Journal of Physics D: Applied Physics, 52*(44), 445303.

Zhu, J., Yan, C., Zhang, X., Yang, C., Jiang, M., & Zhang, X. (2020). A sustainable platform of lignin: From bioresources to materials and their applications in rechargeable batteries and supercapacitors. *Progress in Energy and Combustion Science, 76*, 100788.

Züttel, A. (2003). Materials for hydrogen storage. *Materials Today, 6*(9), 24–33.

Chapter 6

Introduction to Photovoltaic Energy Generation

Roberto Francisco Coelho, Lenon Schmitz,
and Denizar Cruz Martins

Photovoltaic (PV) solar generation is characterized by the direct conversion of the energy contained in solar radiation into electrical energy. A full understanding of this phenomenon requires prior knowledge of the definitions used by experts. In this context, this chapter aims to present the basic topics related to PV energy generation to the beginners, in order to provide to them the conditions for a deep immersion in this theme.

SOLAR ENERGY

The Sun is a star predominantly composed of hydrogen (74%). However, due to the high gravitational force and high temperature to which its nucleus is subjected, its atoms are constantly converted to helium by means of nuclear fusion. In this process, approximately 0.7% of the original mass is converted into energy ($E = mc^2$), implying the emission of photons. As they move inside the Sun, these photons are randomly absorbed and reemitted thousands of times, until finding the solar corona and be expelled to the space in the form of electromagnetic radiation. Before reaching the Earth, the solar radiation travels over 150 million kilometers in the vacuum. Upon entering the Earth's atmosphere, its trajectory is modified due to phenomena such as (United States Department of Energy, 1982):

- Diffusion (or scattering): occurs when particles or gas molecules diffuse the solar radiation, spreading it along the path.
- Absorption: occurs when solar radiation is retained by gas molecules, being converted, as a rule, into thermal energy.

• Reflection: occurs when the solar radiation focuses on reflective barriers, such as the top of clouds and is redirected back to space.

The aforementioned phenomena evidence that the Earth's atmosphere acts as a filter, retaining part of the energy contained in the incident solar radiation. Although being filtered, the portion of the solar radiation that reaches the Earth's surface carries a significant amount of energy and has enormous potential of use. One way to take advantage of this energy is transforming it directly into electrical energy. The devices capable of promoting this conversion are classified as cell, module, or PV array, as detailed below.

Photovoltaic Cell, Module, and Array

PV cells are devices specifically developed to perform the direct conversion of the energy contained in solar radiation into electrical energy. Typically, the voltage generated by a PV cell is about 0.6 V, whereas its current, generally lower than 10 A, is directly proportional to the area exposed to the solar radiation (active area). Structurally, a PV cell is composed of a junction of semiconductors (p-type and n-type) covered by materials that provide mechanical protection, as shown in figure 6.1. To collect the generated current, metallic contacts are welded on both sides of the cell, but the ones on the upper part are strategically built to maximize the active area, allowing the occurrence of the PV effect with higher efficiency.

To lift the photogenerated voltage, PV cells can be electrically connected in series and mechanically encapsulated in rigid structures called PV modules. The output power is limited to few hundreds of watts. If greater powers are still required, PV modules can also be associated in series, parallel, or both to form PV arrays, as shown in figure 6.2 (Coelho and Martins, 2014). It is worth noting that PV arrays formed by modules connected exclusively

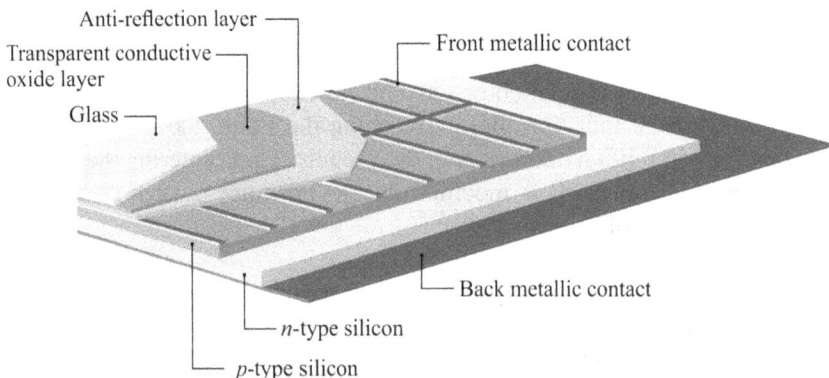

Figure 6.1 **Main Layers of a Typical Solar Cell.** *Source:* Created by the authors.

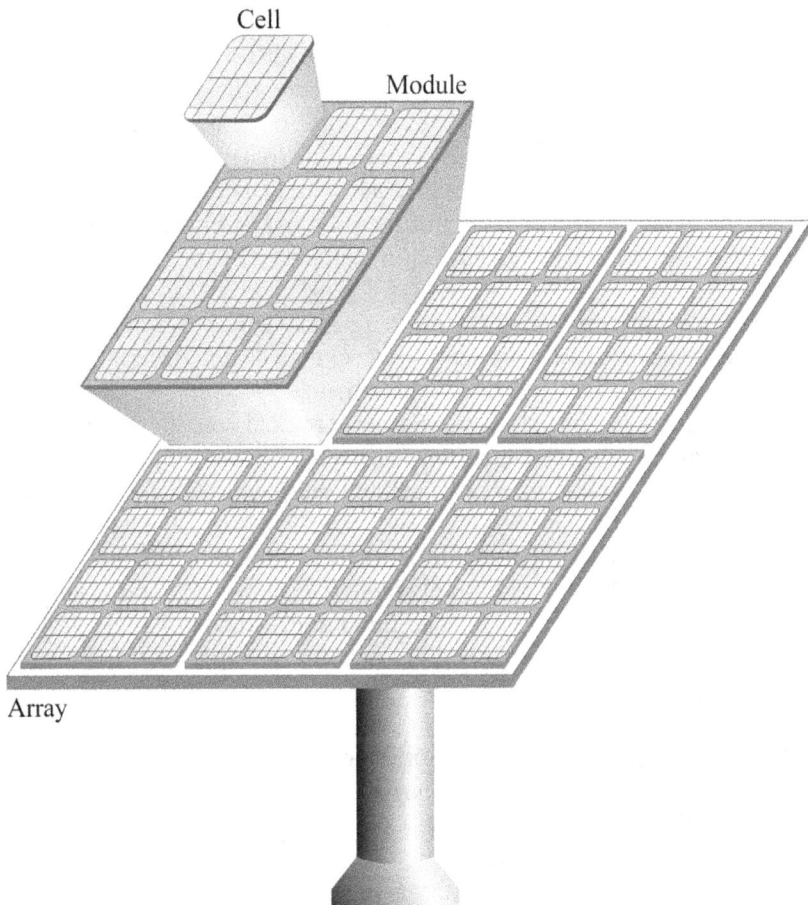

Figure 6.2   Photovoltaic Cell, Module, and Array. *Source*: Created by the authors.

in series are commonly named by PV strings. In general, cells, modules, or arrays are designated as PV generators, a term adopted in this chapter as a literal nomenclature attributed to any device of PV generation.

Types of Photovoltaic Cells

PV cells are usually named according to the materials from which they are manufactured. In general, they can be classified as being of first-, second-, or third-generation (Kibria et al., 2015; Mingsukang et al., 2017):

- First-generation solar cells: These are divided into two productive chains: polycrystalline silicon (p-Si) and monocrystalline silicon (m-Si) cells. Both are considered reliable, have the highest commercially available efficiencies, and hold more than 94% of the PV market (Fraunhofer Institute for Solar Energy Systems, 2020). Even being more efficient than non-silicon based cells, they are more at risk to lose some of their efficiency at higher temperatures (hot sunny days), than thin-film solar cells. In other words, they present higher thermal coefficents (Ruther and Kleiss 1996). Currently, the laboratory efficiencies of the p-Si and m-Si cells are over 20% and 25%, respectively (Fraunhofer Institute for Solar Energy Systems, 2020).

- Second-generation solar cells: These are usually called thin-film solar cells because they are made from layers of semiconductor material with few micrometers of thickness, which makes them less expensive, flexible, and thus advantageous for use on curved surfaces. It is worth mentioning that this technology, which includes the amorphous silicon (a-Si), cadmium telluride (CdTe), and copper indium gallium diselenide (CIGS), still present low penetration in the market because their efficiencies are not competitive when compared to the first-generation cells. The current efficiency of a-Si solar cells has a theoretical limit of about 15%, however, commercial cells have an efficiency lower than 10%. In adition, although CIGS and CdTe solar cells have exceeded the barrier of 20% of efficiency in laboratory, their comercial efficiencies are between 15% and 18% (Fraunhofer Institute for Solar Energy Systems, 2020).

- Third-generation solar cells: These technologies are still emerging and do not have large-scale production. They are potentially able to overcome the Shockley–Queisser limit of 31–41% efficiency for single band gap solar cells. It is worth mentioning that the Shockley–Queisser limit is applied only to single-junction cells. Multi-layer technologies can overcome this barrier. According to Fraunhofer Institute for Solar Energy Systems (2020), the current efficiency of multi-junction cells is 47.1%. There are several technologies classified as third-generation solar cell technologies, which include solar cells sensitized by a dye material, solar cells sensitized by quantum dots (QDs), and perovskite-sensitized solar cells. The first studies on sensitized PV started during the 1970s with the use of organic dyes as the sensitizer, which can be based on natural or synthetic organic dyes. Natural organic dyes can be obtained from plant sources but the performance is poor and the efficiency is low, whereas synthetic organic dyes can give efficiency as high as 13.5% (Fraunhofer Institute for Solar Energy Systems, 2020). The replacement of organic dyes with inorganic sensitizers resulted in the emergence of QD-sensitized solar cells (QDSSCs) that utilize QDs or nanosized semiconductor crystals with a short band gap and a high extinction coefficient. Later, from 2009, the use of perovskite materials as sensitizers become an option. Perovskite works very well with the solid-state hole transfer material and its efficiency in 2019 has reached 21.6% for a cell and 16.1% for a module, according to Fraunhofer Institute for Solar Energy Systems (2020); nevertheless, perovskites are very moisture sensitive materials and fabrication must be done in very clean and controlled conditions (Mingsukang et al., 2017).

Comparison among the Silicon-Based Photovoltaic Cells

As previously stated, most PV cells found in the market today are based on silicon and classified as crystalline silicon, which is subdivided into m-Si and p-Si, or a-Si (Mohammadnoor et al., 2012). Whereas monocrystalline cells have the greatest commercial efficiencies, around 15–20%, polycrystalline cells have efficiencies around 14–18%, and a-Si have around 6–10%. Differences in efficiency levels are intrinsically related to the fabrication process. Monocrystalline-Si cells are made from a single silicon crystal using the Czochralski process, in which a grain of silicon is immersed in molten silicon. During growth, the crystal receives small amounts of boron forming high-purity *p*-type silicon liquid. In this process, the crystal grain is pulled, leading to the formation of a *p*-type silicon ingot which is cut into thin slices and taken to the diffusion furnace, where it receives the doping with phosphorus at high temperatures, thus forming the *p-n* junction.

Polycrystalline-Si cells do not undergo the Czochralski process. In this case, the high-purity *p*-type liquid it is cooled, forming several crystals. These crystals pass by the same doping process with phosphorus, generating a *p-n* junction like m-Si.

Amorphous-Si cells are produced by the deposition of thin silicon films. Amorphous-Si does not form a uniform crystalline network and its disordered structure has many pending connections that form holes. These holes can recombine with free electrons and impair the flow of current through the cell. Therefore, a-Si is hydrogenated, so that the hydrogen atoms occupy the holes, reducing the density of pending bonds and allowing electrons to flow through the cell.

Figure 6.3 highlights the structural differences among the three types of silicon cells, whereas table 6.1 emphasizes their vantages and disadvantages.

Photoelectric and Photovoltaic Effects

The observation that some materials are capable of generating electricity from light was firstly carried out by the physicist Alexandre Edmond Becquerel in 1839. Becquerel, however, failed to explain the origin of such phenomenon, neither the British researchers William Adams and Richard Ray, who built the first rudimentary solid cell with an efficiency of about 0.5%, in 1876.

Roberto Francisco Coelho et al.

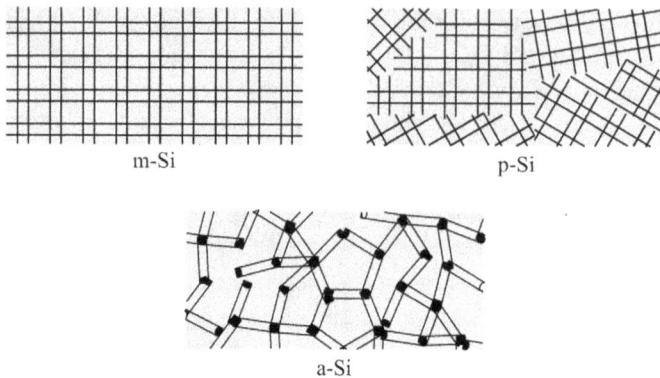

Figure 6.3   **Physical Structure of Photovoltaic Cells Manufactured from Monocristalline (m-Si), Polycristalline (p-Si), and Amorphous (a-Si) Silicon.** *Source*: Created by the authors.

Table 6.1   **Comparison among Monocrystalline, Polycrystalline, and Amorphous Silicon Cells**

Silicon Cells	Advantages	Disadvantages
Monocrystalline	High conversions efficiency, most mature technology, and high reliability	High cost, large consumption and, complex production process
Polycrystalline	Fabricated on cheap substrates and lower cost than monocrystalline	Relatively large silicon consultation and cost, complex production process
Amorphous	Low cost, easiness of mass production, good response to weak light	Low conversion efficiency and low stability

Only in 1905, from the concept of wave-particle duality, Albert Einstein finally explained the photoelectric effect, often confused with the PV effect. Although both phenomena are correlated, they are not the same. In order to understand the differences between them, it is necessary to revisit the atomic model.

Modern Atomic Model

According to the modern atomic model proposed by Niels Bohr, the atom is composed of a central nucleus formed by protons and neutrons and orbited by electrons distributed throughout electron shells, which are identified by the letters *K, L, M, N, O, P,* and *Q.* Despite the number of shells varies

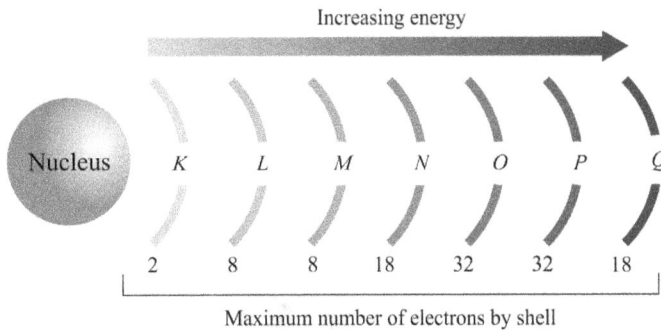

Increasing energy

Nucleus   *K*   *L*   *M*   *N*   *O*   *P*   *Q*

2   8   8   18   32   32   18

Maximum number of electrons by shell

Figure 6.4   Modern Atomic Model: Representation of the Electron Shells of an Atom with Indication of the Maximum Number of Electrons per Shell. *Source*: Created by the authors.

from element to element, the last one is named as valence shell and featured by containing the electrons with the greatest energy level, as illustrated in figure 6.4.

As stated by the octet rule, the atomic stability occurs when the valence shell is filled with eight electrons, because in this condition the atoms acquire similar features as noble gases, becoming stable or inert (nonreactive). To satisfy the octet rule, atoms bond in groups to form stable molecules. Such bonds may occur through donation, reception, or even sharing of electrons, and are classified as ionic, covalent, and metallic bonds, as described below:

- Ionic bonds: These occur through the transfer of electrons from metals (few electrons in the valence shell) to ametals (many electrons in the valence shell), as shown in figure 6.5 (a). When donating electrons, metals stabilize, but become positively charged (cations), whereas upon receiving them, ametals also stabilize, but become negatively charged (anions). Once formed, cations and anions are attracted due to electrostatic forces resulting from the electrical charges acquired during the process.
- Covalent bonds: These occur through the sharing of electron pairs between adjacent atoms of semiconductor materials, as shown in figure 6.5 (b). For example, two atoms with four electrons in the valence shell may share these electrons so that together they have eight electrons in the valence shell.
- Metallic bonds: These come from the cohesion between atoms of metallic materials, characterized by the electropositivity, that is, by the tendance of donating electrons. The force that holds these atoms together cannot be explained by the octet rule, which is replaced by the electron-sea model, stated as follows: along the structure of the metal, free electrons constantly detached from their respective nucleus, causing the atoms, previously

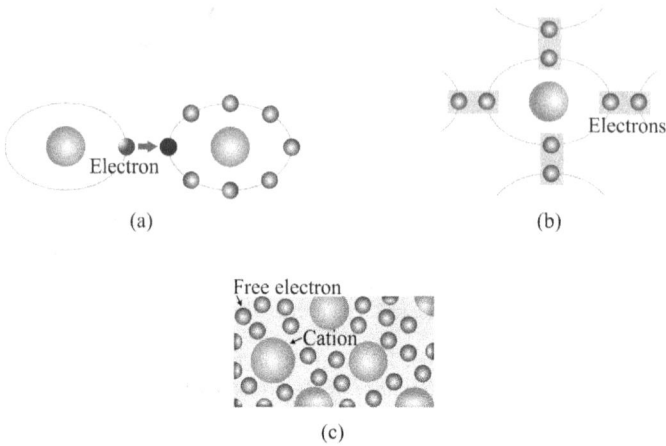

Figure 6.5   Atomic Bonds: (a) Ionic Bond between Metal and Ametal; (b) Covalent Bond between Semimetals, and (c) Metallic Bond between Metallic Materials. *Source*: Created by the authors.

electrically neutral, to become cations inside a sea of electrons. Figure 6.5 (c) illustrates the representation of this phenomenon.

Photoelectric Effect

The photoelectric effect is featured by the emission of free electrons from metallic materials exposed to electromagnetic radiation (photons), as shown in figure 6.6 (Arons and Peppard, 1965). The equation that describes this phenomenon can be obtained by applying the energy conservation law to the photon-material interaction, which allows writing:

$$E_k = E_{ph} - W \tag{6.1}$$

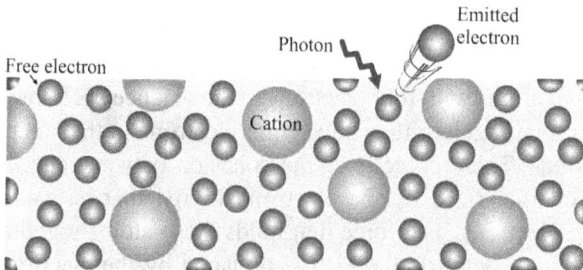

Figure 6.6   Representation of the Photoelectric Effect Occurrence. *Source*: Created by the authors.

where, E_k is the kinetic energy of the emitted electrons, W is the minimum energy that each electron must absorb to escape from the material (work function), and E_{ph} is the energy of a photon, calculated by:

$$E_{ph} = hf. \tag{6.2}$$

In this last equation, f is the frequency of the incident photon and $h = 4.14 \cdot 10^{-15}$ eVs is the Planck constant. The occurrence of the photoelectric effect is conditioned to the incidence of photons with enough energy to release electrons from the material. This condition can be mathematically translated into:

$$E_{ph} > W. \tag{6.3}$$

Any additional energy transported by the photon before the collision is converted into kinetic energy after the collision. Thus, the escape velocity v_e of the emitted electron can be obtained according to:

$$E_k = \frac{1}{2}m_e v_e^2 \rightarrow v_e = \sqrt{\frac{2E_k}{m_e}} = \sqrt{\frac{2\left(E_{ph} - W\right)}{m_e}} = \sqrt{\frac{2\left(hf - W\right)}{m_e}}, \tag{6.4}$$

where, m_e is the electron rest mass.

Photovoltaic Effect

The PV effect occurs only in semiconductors and, therefore, the Si, a material abundant on the Earth's surface, is widely used in the PV cell manufacturing process. (United States Department of Energy, 1982) Pure Si atoms have four electrons in the valence shell, as shown in figure 6.7 (a). However, according to the octet rule, its atomic stability is only achieved when its valence shell is filled with eight electrons. Thus, seeking to stabilize, the Si atoms share electrons through covalent bonds, giving rise to crystalline structures, as depicted in figure 6.7 (b).

A semiconductor crystal made from pure silicon is named an intrinsic semiconductor, since it does not contain atoms of other materials. Like other solids, when exposed to light (photons), the intrinsic silicon crystal absorbs part of the incident energy, but if the radiation is of low energy (low frequency), there are no changes in its electrical properties, only heating. This is because the absorption of low-energy radiation increases the thermal agitation of the crystal atoms. In addition, under such a condition, the electrons of the electrosphere are excited and jump to more energetic shells; however,

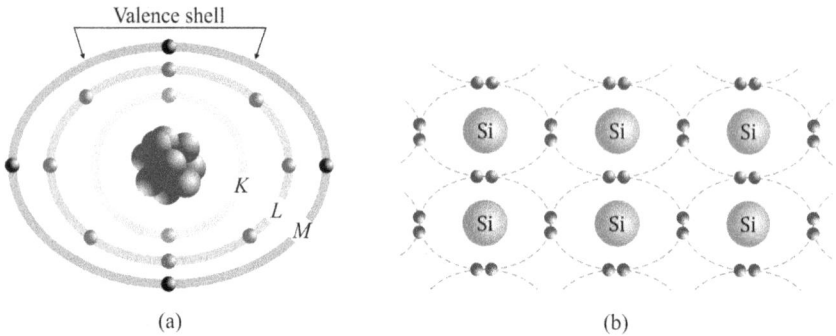

Figure 6.7 **Silicon Atom: (a) Representation of the Silicon Atom; (b) Representation of the Covalent Bonds among Atoms of a Pure Silicon Crystal.** *Source*: Created by the authors.

as this condition tends to be unstable, they return to the initial state and the absorbed energy is also converted into heat (Fahrenbruch and Bube, 1983).

On the contrary, if the absorbed radiation contains enough energy to release the electrons from the atomic nucleus and make them free, there are changes in the electrical properties of the crystal, as each electron that passes to the conduction band leaves an empty space in the valence shell. These empty spaces are called holes, as illustrated in figure 6.8 (Fahrenbruch and Bube, 1983).

It is worth mentioning that the appearance of free electrons does not result in the circulation of electrical current, because free electrons and holes emerge in pairs and present high probability of recombination. In

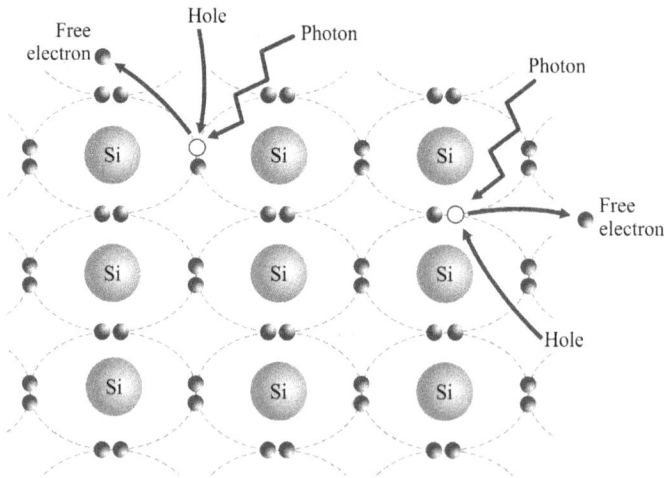

Figure 6.8 **Generation of Electron-Hole Pairs in a Silicon Crystal Exposed to High-Energy Radiation.** *Source*: Created by the authors.

other words, when excited by photons, free electrons migrate to the conduction band, but when returning to the valence band, they recombine with one of the holes generated just before. Thus, the occurrence of electric current in the silicon crystals is conditioned to the existence of a force that accelerates the free electrons from the conduction band to an external electrical circuit, preventing them from returning to the valence shell after being excited. Such condition can be achieved by applying a permanent electric field, generated by the proper doping of each of the silicon substrates that form the *p-n* junction (Green, 1982).

The doping process consists of the purposeful addition of impurities to the intrinsic semiconductor, in order to increase the amount of free electrons and holes. Impurity atoms can be introduced into the crystal by pressure, occupying the adjacent spaces between the silicon atoms, or in the replacement of some original silicon atoms. In both cases, the crystal structure is modified.

On doping the silicon with phosphorus (P) atoms, said to be pentavalentes because they contain five electrons in the valence shell, a crystal with excess electrons called *n*-type silicon (negative charge carriers) is obtained. In this material, the four electrons of the valence shell of the silicon atoms make covalent bonds with four of the five electrons of the valence shell of the phosphorus atoms, remaining electrons weakly attached to the crystalline structure of the semiconductor as shown in figure 6.9,[1] parts (a) and (b) (Moller 1993).

When repeating the process, but considering boron (B) atoms, said to be trivalent because they have three electrons in the valence shell, a crystal with excess holes called *p*-silicon (positive charge carriers) is obtained. In this case, three of the four electrons of the valence shell of the silicon atoms make covalent bonds with the three electrons of the valence shell of the boron atoms, remaining holes to be filled in the crystal structure as per figure 6.9, parts (c) and (d) (Moller 1993).

Although *n*-type crystals have excess electrons and *p*-type crystals have excess holes, as highlighted in figure 6.10 (a), both are electrically neutral when separated, because the net charge (sum of the charges of electrons and protons) of each of the crystals (*n* and *p*), even after doping, remains zero. However, when such crystals are joined, free electrons of the *n*-side migrate to fill the holes on the *p*-side and vice versa. This transfer of charge carriers occurs quickly throughout the entire border between the two materials as shown in figure 6.10 (b). When free electrons diffuse to the *p*-side, they find holes in excess to be filled. Likewise, when the holes diffuse to the *n*-side, they find free electrons to be captured. This process results in the depletion of charge carriers in the vicinity of the *p-n* junction, which now contains only fixed charges, as depicted in figure 6.10 (c).

It is worth mentioning that before the diffusion of charge carriers, the *n*-type crystal has an excess of electrons; nevertheless, it is electrically neutral, since the negative charge of its electrons is compensated by the positive charge of its protons. When donating electrons to the p-side, the Si-P molecules on the n-side become positively charged, but still with eight electrons in the valence shell and, thus, inert. Similarly, the Si-B molecules on the p-side are electrically neutral before the diffusion and have an excess of holes in the crystal lattice. However, upon receiving electrons from the n-side, they become negative fixed charges with eight electrons in the valence shell (United States Department of Energy, 1982).

Despite it is intuitive to believe that the diffusion of electrons and holes occurs indefinitely until the depletion layer deepens in the entire crystal, it does not happen, because as the fixed charges are formed, they impose an electric field contrary to the flow of the charge carriers. Therefore as the charge carriers cross the p-n junction, the greater is the concentration of fixed charges in their surroundings, more intense is the electric field making diffusion a self-limited process as demonstrated in figure 6.11 (a) (Green 1982; United States Department of Energy 1982).

After the electric field is established, the electrons of the n-side stop migrating to the p-side; likewise, the holes of the p-side are no longer able to migrate to the n-side. Conversely, any hole that appears on the n-side is accelerated by the electric field to the p-side, as any electron that appears on the p-side is accelerated to the n-side. Therefore, the electric field acts to separate the charge carriers, maintaining the electrons on the n-side and the holes on the p-side and hence generating a potential difference (V) between both substrates, as per figure 6.11 (b).

Now, if a photon with enough energy to release an electron reaches the p-side of the crystal, as in figure 6.12 (a), an electron-hole pair is created. The electron remains free only for a short time interval, due to the high probability of recombination with one of the excess holes of the p-side. During this short interval, however, the electron describes a random path within the crystal before entering in the p-n junction and being accelerated by the electric field to the n-side, whereas hole remains confined on the p-side. If the photon reached the n-side, a similar process would be established, but in this case, the electron would be confined to the n-side and a hole would be accelerated by the electric field to the p-side. Therefore, electrons generated on the p-side are accelerated by the electric field to the n-side, and holes generated on the n-side are accelerated to the p-side (Smets et al., 2016).

Finally, when an external circuit is connected between the terminals of the p-n junction, hereinafter referred to as PV cell, the excess electrons on the n-side flow to the p-side (through the external circuit), where the recombination occurs, as depicted in figure 6.12 (b). Under abundant solar radiation, new pairs of electrons-holes are constantly created and separated by the action of the electric field, ensuring the continuity of the current through the external circuit and the permanent occurrence of the PV effect (Green, 1982).

Conditions for the Occurrence of the Photovoltaic Effect

Based on the basic concepts previously presented, it is possible to conclude that the occurrence of the PV effect requires (United States Department of Energy, 1982):

- Existence of charge carriers: doping process.
- Excitation of charge carriers by an external source: photoelectric effect.
- Separation of charge carriers: electric field in the *p-n* junction.

In the case of a semiconductor, the energy contained in the incident photon must be higher than the band gap energy (E_G), which is the minimum energy that an electron must absorb to migrate from the valence to the conduction band. Thus:

$$E_{ph} > E_G \rightarrow hf > E_G. \tag{6.5}$$

The band gap energy is dependent on the temperature and can be rigorously approximated by:

$$E_G = E_{G0} - \frac{K_1 T^2}{T + K_2}, \tag{6.6}$$

where, E_{G0}, K_1, and K_2 are constants defined for each semiconductor material, as listed in table 6.2, and T is the temperature in Kelvin.

On applying the values of table 6.2 in equation (6.6), considering a silicon PV cell subjected to a temperature $T = 25°C$ ($T = 298$ K), one can obtain a bandgap energy of about 1.12 eV. Thus, replacing this value into equation (6.5), the minimum frequency (cut-off frequency) that guarantees the occurrence of the PV effect can be determined:

$$hf > E_G \rightarrow f > \frac{E_G}{h} = \frac{1.12}{4.14 \cdot 10^{-15}} = 2.71 \cdot 10^{14} \text{ Hz}. \tag{6.7}$$

Equation (6.7) reveals that the cut-off frequency is located at the upper limit of the infrared spectrum; therefore, any electromagnetic wave with

Table 6.2 Empirical Constants for Silicon and Germanium Semiconductors

	Silicon	Germanium
E_{G0}	1.166 eV	0.744 eV
K_1	$4.73 \cdot 10^{-4}$ eV	$4.77 \cdot 10^{-4}$ eV
K_2	636 K	235 K

frequency above this level contains enough energy to guarantee the occurrence of the PV effect in a silicon cell, as evidenced by figure 6.13.

MAIN VARIABLES INVOLVED IN
THE PHOTOGENERATION

The operation of PV generators is directly dependent on two weather factors: solar radiation and temperature, which affect the voltage, current, and power levels delivered by the PV generator. Factors such as orientation and inclination also affect the photogeneration.

Temperature

The temperature T [K] is a factor directly associated with the thermal ionization of the charge carriers on the semiconductor crystal, impacting on the recombination process and, thus, on the levels of the generated voltage (Moller, 1993).

Solar Radiation and Solar Irradiance

The energy emanated by the Sun, known as solar radiation (S) and measured in joules [J], reaches the Earth's surface through electromagnetic waves that are propagated in a vacuum at the speed of light. One way to quantify the solar radiation is through the solar irradiance (G), which is the power contained in the radiation that reaches a square meter of a given surface, being measured in watt per square meter [W/m²] (Coelho and Martins, 2014).

According to the World Meteorological Organization, the outer part of the Earth's atmosphere is exposed to an average irradiance of 1,367 W/m². However, due to phenomena such as diffusion, reflection, and absorption, only a part of this total energy reaches the Earth's surface, as illustrated in figure 6.14. Although on sunny days the solar irradiance on the Earth's surface is expressive, in the order of 1,000 W/m², this value varies according to the latitude and cloudiness, as indicated in figure 6.15.

In view of the aforementioned features, one can conclude that the Earth's atmosphere behaves as a filter for the solar radiation; nevertheless, the filtering effect is nonlinear, since the molecules present in the air, especially ozone (O_3), oxygen (O_2), and water (H_2O), act at specific wavelengths. While ozone attenuates the components with wavelengths close to 250 nm (ultraviolet), oxygen occasionally attenuates the components with wavelength close to 750 nm, and the water molecules cause attenuations at 900, 1,150, 1,350, 1,850, and 2,500 nm. As a result, the attenuation becomes more pronounced at some wavelengths.

One way to quantify these filtering effects is determining the thickness of the atmosphere layer in the radiation path. For this purpose, the concept of air mass index, or simply Air Mass (AM), is used which relates the path L traveled by radiation until reaching a certain point on the Earth's surface in relation to the shortest possible path L_0 as shown in figure 6.16 (Coelho and Martins 2014). The AM index may be mathematically approximated (this approximation disregards the curvature of the Earth, being reasonably accurate for θ_z lower than 75°) by:

$$AM = \frac{L}{L_0} = \frac{1}{\cos(\theta_z)}. \tag{6.8}$$

On Earth's surface, the AM index varies according to the azimuthal angle θ_z, measured in relation to the line normal to the plane that tangency the point of interest. The greater is θ_z, greater is the thickness of the air layer in the radiation path, and lower is the energy that reaches this point. As there is no air outside Earth's atmosphere, there is no attenuation at any wavelength. In this condition, the AM index is called AM_0.

Orientation and Inclination

It is important to highlight that regardless of the thickness of the air layer in the radiation path, the photogenerated energy is maximized when the direct component of the irradiance strikes perpendicularly to the surface of the PV generator. Since the position of the Sun in relation to a point on the Earth's surface varies throughout the day and the seasons, the maximization is conditioned to the use of so-called solar trackers. These tracking mechanisms have the purpose of making the PV generators follow the Sun as it moves across the sky.

Although this solution is ideal, the usage of trackers implies high costs, the need for more space, and maintenance due to the existence of moving parts. In addition, as part of the energy generated is used to drive the engines, feasibility studies that prove the energy gain also need to be carried out. For these reasons, in most applications the PV generators are installed in a fixed position and, thus, they should be properly oriented and tilted to ensure the maximization of the average energy annually generated (Ahmad et al., 2003).

From the orientation point of view, the best solar incidence occurs when the PV generators are oriented toward the equator; therefore, they should face the geographical south when installed in the northern hemisphere and toward the geographical north when installed in the southern hemisphere, as shown in figure 6.17 (a). Such characteristics are related to the trajectory described by the Sun. In the northern hemisphere, it rises in the east, culminates in

Table 6.3   **Recommended Tilt Angle Depending on the Latitude of the Installation Site**

Location Latitude	Recommended Tilt Angle
0°–10°	10°
11°–20°	Location latitude
21°–30°	Location latitude + 5°
31°–40°	Location latitude + 10°
41° or more	Location latitude + 15°

the south while moving to the right, and sets in the west. By contrast, in the southern hemisphere, it rises in the east, culminates in the north while moving to the left, and sets in the west, as illustrated in figure 6.17 (b).

Noticeably, as PV generators are usually installed on roofs already built, it is not always possible to guarantee perfect alignment. In these cases, the orientation to the northeast or northwest is also acceptable, with losses between 3% and 8%. Conversely, when the installation is carried out facing east or west, the losses can reach even higher levels, between 12% and 20%. The tilt angle is another important factor to be considered, since it determines how the Sun rays strike the surface of the PV generator. In order to maximize the energy generated over a year, studies indicate that the best slope occurs when the angle is established close to the latitude of the location, being optimized for each region if the recommendations in table 6.3 are adopted (Kyocera Installation Manual, 2009).

STANDARDIZATION OF WEATHER CONDITIONS AND TESTS FOR ELECTRICAL CHARACTERIZATION

The standardization of the climatic conditions adopted to perform the electrical characterization of the PV generators ensures the uniformity of the information contained in the technical datasheets and allows a direct comparison among modules from different manufacturers (Coelho and Martins, 2014).

The first set of specifications adopted by the PV industry is called Standard Test Conditions (STC) and defines the values of solar irradiance, temperature, and AM index to be $G = 1000$ W / m^2, $T = 25$ °C, and $AM = 1.5$, respectively. Such definitions are not consistent with typical real-world operating conditions, since under 1,000 W/m^2 irradiance it is unlikely that the PV generator temperature will settle at just 25°C. Therefore, it is improbable that a generator installed in the field will supply the power reported in the technical sheet.

To overcome this problem, many manufacturers also electrically characterize their PV generators in weather conditions closer to those found in the field. For this purpose, they use irradiance of $G = 800$ W/m^2 and AM coefficient

Table 6.4 Comparison among the Weather Variables Defined in Standard Test Conditions and Nominal Operating Cell Temperature

	Irradiance	*Air Mass*	*Cell Temperature*
STC	1000 W/m²	1.5	25°C
NOTC	800 W/m²	1.5	45 ± 3°C

of 1.5, considering a nominal operating cell temperature (NOCT), which is typically set at 45 ± 3°C. It is worth noting that the NOCT is determined experimentally by the manufacturers, and its value is listed in datasheets. To measure it, the PV generator is maintained at open circuit and subjected to irradiance of $G = 800$ W/m², AM coefficient of 1.5, ambient temperature of 20°C, and wind speed of 1 m/s. Table 6.4 differentiates the weather conditions adopted in both, STC and NOCT (Lasnier and Ang, 1990).

Characteristics Curves and Figures of Merit

The electrical quantities that characterize PV generators are synthesized in the form of curves that relate current to voltage (I-V curve) or power to voltage (P-V curve). Regardless of the PV module analyzed, these curves have the typical shape depicted in figure 6.18 and, for this reason, are known as characteristic I-V and P-V curves. The main points related to these curves are (Lasnier and Ang, 1990):

- Maximum power voltage (V_{mp}): it is the value of the photogenerated voltage when the PV generator operates at the maximum power point (MPP).
- Maximum power current (I_{mp}): it is the value of the photogenerated current when the PV generator operates at the MPP.
- Maximum power (P_{mp}): refers to the maximum generated power, calculated by the product between V_{mp} and I_{mp}. When specified in the STC, the maximum power is expressed in peak watt [W_p].
- Open-circuit voltage (V_{oc}): it is the value of voltage when the terminals of the PV generator are kept open. In this situation, the photogenerated current and power are zero.
- Short-circuit current (I_{sc}): it is the value of current when the terminals of the PV generator are kept in short circuit. In this condition, the photogenerated voltage and power are zero.

In addition to the variables V_{mp}, I_{mp}, P_{mp}, V_{oc}, and I_{sc}, there are other figures of merit related to the I-V and P-V curves that serve as metrics for identifying the quality of a PV generator, such as the fill factor (FF) and the maximum efficiency.

Fill Factor

The fill factor (FF) is defined as the ratio between the areas calculated by the products $V_{mp}I_{mp}$ and $V_{oc}I_{sc}$, as represented in figure 6.19. It is mathematically described by (Lasnier and Ang, 1990):

$$FF = \frac{V_{mp}I_{mp}}{V_{oc}I_{sc}} = \frac{P_{mp}}{V_{oc}I_{sc}}. \qquad (6.9)$$

The higher the FF, the better the performance of the PV generator. However, as the voltage at the MPP is lower than the open-circuit voltage and the current at the MPP is lower than the short-circuit current, the FF will always be less than the unit, normally settling between 0.7 and 0.85.

Maximum Efficiency

The maximum efficiency (η_{max}) of the photoconversion is given by the ratio between the maximum electrical power provided by the PV generator (STC) and the power contained in the incident solar radiation that reaches its surface area A_{PV}, being calculated by (Lasnier and Ang, 1990):

$$\eta_{max} = \frac{P_{mp}}{A_{PV} \cdot G} 100\%. \qquad (6.10)$$

No matter how sophisticated is the manufacturing process; the maximum efficiency of PV cells in the same batch can vary by up to ± 5% in relation to the nominal value. That is why many manufacturers offer modules with very close power ratings, such as the Kyocera KD200GX series, whose modules have powers of 240, 245, and 250 W.

Influence of the Irradiance and Temperature
on the I-V and P-V Curves

Unfortunately, environmental factors over which there is no control, such as solar irradiance and temperature, significantly affect the photogenerated voltage and current levels. As a consequence, the FF and the maximum efficiency cannot be assumed as constant parameters, since they vary depending on the environmental conditions of the installation site (Villalva et al., 2009). In this sense, the characterization of PV generators is described not only by a curve, but by a family of curves that account for different irradiance and temperature scenarios, as shown in the manufacturers' datasheets and in the illustration of figure 6.20.

In order to understand the behavior of these curves, the reader should keep in mind that when the irradiance on a PV generator varies, the number of photons per second that passes through it also varies. Since each photon interacts with an electron, releasing it from the semiconductor structure after the collision, there is a proportional relationship between the number of incident photons and the number of electrons that become free. In other words, there is a linear proportion between the solar irradiance and the intensity of the photogenerated current. In addition, when the operating temperature rises, the atoms of the semiconductor crystal vibrate more intensely, a fact that reduces the bandgap energy and results in the emission of a greater number of electrons, with a slight increase in photogenerated current (United States Department of Energy, 1982).

Mathematically, such effects are felt in the short-circuit current, which value can be calculated by:

$$I_{sc} = \frac{G}{G^{STC}} I_{sc}^{STC} \left[1 + \alpha \left(T - T^{STC} \right) \right], \tag{6.11}$$

where, G and T are the irradiance and temperature values in which the generator is subjected, G^{STC} and T^{STC} are their respective values at the STC (1000 W/m^2 and 25°C), I_{sc}^{STC} is the short-circuit current at the STC, and α [%/°C] is the thermal coefficient that describes the ratio between the short-circuit current and the temperature:

$$\alpha = \frac{1}{I_{sc}^{STC}} \frac{\Delta I_{sc}}{\Delta T}. \tag{6.12}$$

Due to the increase in temperature, some free electrons and holes also gain thermal energy to diffuse in the opposite direction to that indicated by the electric field of the p-n junction, implying a reduction in the photogenerated voltage due to the narrowing of the depletion layer. Mathematically, the equation that describes the behavior of the open-circuit voltage as a function of the temperature is written as:

$$V_{oc} = V_{oc}^{STC} \left[1 + \beta \left(T - T^{STC} \right) \right], \tag{6.13}$$

in which V_{oc}^{STC} is the open-circuit voltage at the STC and β [%/°C] is the thermal coefficient that represents the ratio between the open-circuit voltage and the temperature:

$$\beta = \frac{1}{V_{oc}^{STC}} \frac{\Delta V_{oc}}{\Delta T}. \tag{6.14}$$

Because they affect voltage and current, variations in irradiance and temperature also become noticeable in the photogenerated power, whose behavior is described by:

$$P_{mp} = \frac{G}{G^{STC}} P_{mp}^{STC} \left[1 + \gamma \left(T - T^{STC} \right) \right],$$

(6.15)

so that P_{mp}^{STC} is the maximum power at the STC and γ [%/°C] is the thermal coefficient that represents the ratio between the maximum power and the temperature:

$$\gamma = \frac{1}{P_{mp}^{STC}} \frac{\Delta P_{mp}}{\Delta T}.$$

(6.16)

In general, manufacturers provide the thermal coefficients α, β, and γ related to I_{sc}, V_{oc}, and P_{mp}, respectively, but they do not refer to the thermal coefficients of V_{mp} and I_{mp}. Thus, when these coefficients are required, it is necessary to estimate them. First, it is assumed that the currents I_{sc} and I_{mp} are subject to the same thermal coefficient, so that one can write:

$$I_{mp} = \frac{G}{G^{STC}} I_{mp}^{STC} \left[1 + \alpha_{mp} \left(T - T^{STC} \right) \right],$$

(6.17)

where I_{mp}^{STC} represents the current at the maximum power point and α_{mp} [%/°C] is the thermal coefficient that quantifies its ratio as a function of temperature:

$$\alpha_{mp} \approx \alpha = \frac{1}{I_{sc}^{STC}} \frac{\Delta I_{mp}}{\Delta T}.$$

(6.18)

In addition, in view of the linear behavior of the open-circuit voltage described in equation (6.12), it is reasonable to assume that V_{mp} also varies linearly with temperature. Therefore:

$$V_{mp} = V_{mp}^{STC} \left[1 + \beta_{mp} \left(T - T^{STC} \right) \right],$$

(6.19)

being V_{mp}^{STC} the voltage at the maximum power point and β_{mp} [%/°C] the thermal coefficient that represents the ratio between the voltage at the maximum power point and the temperature:

$$\beta_{mp} = \frac{1}{V_{mp}^{STC}} \frac{\Delta V_{mp}}{\Delta T}.$$

(6.20)

Table 6.5 Thermal Coefficients of Some Commercial Photovoltaic Modules

Manufacture	Module	α [%/°C]	β [%/°C]	γ [%/°C]
Mitsubishi	MLT265-HC	0.056	-0.350	-0.450
Suntech	STP250S-20	0.050	-0.340	-0.450
Sharp	ND-250QCS	0.053	-0.360	-0.485
Kyocera	KD250GX	0.060	-0.360	-0.460

Now, considering that the expression $P_{mp} = V_{mp}I_{mp}$ is always true, one can use equations (6.17) and (6.19) to obtain:

$$P_{mp} = \frac{G}{G^{STC}} V_{mp}^{STC} I_{mp}^{STC} \left[1 + \left(\alpha_{mp} + \beta_{mp} \right)\left(T - T^{STC} \right) + \beta_{mp}\alpha_{mp} \left(T - T^{STC} \right)^2 \right].$$

(6.21)

The evaluation of the thermal coefficients extracted from datasheets of some manufacturers, described in table 6.5, shows that β and γ are much greater than α. This observation allows us supposing that β_{mp} is also significantly greater than α_{mp}, thus, since both β_{mp} and α_{mp} are much lower than the unity the quadratic term of equation (6.21) becomes negligible compared to its others. Thus, equation (6.21) can be simplified as:

$$P_{mp}(T) = \frac{G}{G^{STC}} P_{mp}^{STC} \left[1 + \left(\alpha_{mp} + \beta_{mp} \right)\left(T - T^{STC} \right) \right].$$

(6.22)

When comparing equations (6.15) and (6.22) it appears that:

$$\gamma = \alpha_{mp} + \beta_{mp}.$$

(6.23)

Furthermore, because α_{mp} can be considered equal to α, one can write:

$$\beta_{mp} = \gamma - \alpha.$$

(6.24)

It should be mentioned that some authors consider that $\beta_{mp} = \gamma$, since $|\gamma| \gg |\alpha|$.

The set of equations composed by (6.11), (6.13), (6.15), (6.17), and (6.19) allows correcting the output quantities of a PV generator for irradiance and temperature values different from those established at the STC. If it is desired to evaluate only the impacts caused by temperature variations, it should be assumed that $G = G^{STC}$. Analogously, in the condition that $T = T^{STC}$, only the effects caused by variations in the irradiance is considered.

Basically, irradiance and temperature affect the output voltage and current of PV generators in different proportions. Variations in irradiance predominantly alter the current, but have a neglectable influence on the photogenerated voltage. Conversely, temperature variations are related to changes in the voltage levels, but do not significantly change the photogenerated current.

External Factors Leading to Reduction of Photogenerated Power

The reduction in the maximum power provided by crystalline silicon PV modules, in relation to the value established in laboratory at the STC, occurs since the first hours of exposure to solar radiation due to the effect called light-induced degradation (LID) (Alexander et al., 2014). Simply put, LID is caused by the existence of traces of oxygen in the molten silicon during the Czochralski process to obtain the crystals that will give rise to the PV cells. When the crystalline modules are first exposed to light, the oxygen molecules associate with boron atoms (*p*-type doping material) form structures capable of capturing electrons and holes, implying a reduction in free charge carriers during the occurrence of the PV effect. Because it is associated with the purity of the semiconductor material, LID depends on the sophistication of the process used in the production of silicon crystals. In most cases, the reduction in the generated power due to this phenomenon is between 1% and 3% of the value specified at the STC.

After the first exposure to light, the degradation of the modules occurs approximately linearly. Reports presented by the National Renewable Energy Laboratory (Osterwald et al. 2002) examined the performance of different modules over several years submitted to real weather conditions, and concluded that the typical rate of degradation of crystalline silicon modules is 0.5% per year for modules manufactured before 2000, and 0.4% for those manufactured after this year. Logically, it is the installation conditions and the weather characteristics that establish together the real rate of degradation. Frequently, modules installed in regions with extreme weather (with severe radiation, snow or wind conditions) degrade more rapidly (about 1% per year) than those installed in regions with mild climate (about 0.2% per year).

Although there are several agents that cause long-term degradation, the most impacting factors are mechanical and optical, as detailed below:

• Corrosion: it is caused by the diffusion of water vapor in the package that, in the long run, results in the deterioration of electrical contacts.
• Thermomechanical stress: it arises from the alternation between day and night. This effect results in cracks in the contact surfaces between materials with different expansion coefficients.

- Photodegradation and thermodegradation: both effects cause loss of elasticity, optical transmission, and insulation problems. They are associated with the absorption of ultraviolet radiation, which over time causes deterioration of the mechanical and physical properties of the material.
- Mechanical stress (static and dynamic): it is caused by external forces, such as wind and snow.

Because the degradation is pronounced by the reduction of the generation capacity, it is expected that the levels of the photogenerated voltage and current are also reduced. Recently, some studies have been conducted to quantify the percentage of reduction of these quantities individually. They have pointed to the fact that the main cause of aging is associated with the absorption of ultraviolet radiation, making the cells gradually opaque. In this sense, the reduction in power is inherently linked to the decrease in the photogenerated current, given that the greater opacity of the cells filters the incident radiation, reducing the number of photons that reach the p-n junction and, consequently, the number of free electrons.

The information regarding the percentage of degradation dictates the rules that manufacturers use to determine the warranty of their modules. Currently, most manufacturers guarantee that the peak power of their PV generators will not fall below 90% of the value specified at the STC in the first ten years, neither below 80% in twenty-five years. It is worth mentioning these percentages of generations can only be guaranteed if the PV modules are correctly installed (orientation and inclination), clean and not shaded.

Although it is intuitive to think that the exposure of generators to rain is sufficient to promote cleaning, only non-impregnated and recent residues are removed in this way, which makes necessary to perform periodic cleaning, especially in regions with low levels of precipitation. In practical terms, some studies indicate that a thin layer of dust deposited on the surface of the PV generator can reduce its generation capacity by up to 5%. Nevertheless, opaque pollutants, such as sap, bird droppings, leaves, and branches that fall and adhere to the surface of a PV generator, can completely block the incidence of solar radiation, resulting in losses of up to 25% of the generation capacity. In extreme cases, the accumulation of dirt can lead to the proliferation of fungi that cause stains or even corrosion of its surface, resulting in irreversible damage and permanent reduction of the generation capacity.

It is worth noting that the pattern in which the residue is distributed on the generator surface leads to different types of losses. The uniform distribution of dust, as in figure 6.21 (a), for example, results in an increase on the PV module apparent opacity: the dust layer acts as a filter for incident radiation, reducing the energy that reaches the PV surface. By contrast, in cases where the distribution of dirt is not uniform, as in figure 6.21 (b), the reduction in the power occurs mainly due to the mismatch of voltage and current between clean and dirty cells. In this case, the accumulation of dense dirt on part of one of the cells of a PV module, for example, will reduce its active area and, consequently, its capacity of generating current.

The complete blockage of solar radiation can also be caused by partial shading, which usually occurs in PV installations that contain a large number of modules. In this case, part of the arrangement remains shaded for a time interval, which can extend from seconds to hours. Its effect is similar to that caused by the accumulation of dense dirt on the generator surface, since again there will be a reduction in the active area.

It is also important to realize that shading is predictable (caused by trees, posts, buildings in the surroundings, or when one module shadows the other) or unpredictable (when tree leaves or bird droppings are deposited on the modules), negative impacts include the appearance of hot spots on the PV modules surface and the consequent reduction of the photogenerated power.

SUMMARY

This section presents a summary of the main topics addressed in this chapter and has the objective to provide a quick review to the readers.

- The Earth's atmosphere acts as a filter for the solar radiation. At the top of the atmosphere the irradiance is approximately 1367 W/m^2, but due to the phenomena such as diffusion, absorption, and reflection, this value is generally less than 1000 W/m^2 on the Earth's surface.
- PV cells are the primary devices of PV generation. The cells association allows the construction of PV modules, as well as the association of modules results in PV arrays. Generically, cells, modules, and arrays are called as PV generators.
- Currently, PV cells are differentiated into three generations. The first contemplates crystalline silicon cells, the second consists in those made from thin films and, in contrast, the third describes the emerging cells, which are promising but not yet manufactured on commercial scale.
- The photoelectric and PV effects are different. While the photoelectric effect is associated with the emission of electrons by metallic surfaces, the PV effect is associated with the appearance of a potential difference in properly doped semiconductor materials.
- The power provided by a PV generator is strongly influenced by weather variables under which there is no control, such as solar irradiance G [W/m^2] and temperature T [K].

- The photogenerated current is directly related to the solar irradiance, as well as the photogenerated voltage is directly related to the temperature.
- PV generators are electrically characterized under controlled environmental conditions of irradiance, temperature, and air mass index, respectively given by: $G = 1000$ W/m^2, $T = 25$°C, and $AM = 1.5$. Such specifications dictate STC and are provided in manufacturers' datasheets.
- Due to variations in the generated voltage and current because of the environmental conditions, manufacturers electrically characterize the PV modules by presenting families of curves, called I-V and P-V characteristic curves.
- In order to compare modules from different manufacturers, figures of merits such as fill factor and maximum efficiency (η_{max}) are commonly used.
- The linear thermal coefficients α, β, and γ describe the rate at which the short-circuit current, open-circuit voltage, and maximum power vary with temperature.
- The correction of the generated voltage, current, and power values as a function of the levels of solar irradiance and temperature can be performed by applying equations (6.11), (6.13), (6.15), (6.17), and (6.19).
- Effects associated with aging degradation, accumulation of dirt, and/or partial shading, as well as poor inclination or orientation, can drastically reduce the power provided by a PV generator, even under favorable environmental conditions.

NOTES

1. Figures 6.9 through 6.21 available online at https://rowman.com/ISBN/9781793625021/Sustainable-Engineering-for-Life-Tomorrow.

REFERENCES

Ahmad, G. E., H. M. S. Hussein, and H. H. El-Ghetany. "Theoretical analysis and experimental verification of PV modules." *Renewable Energy*, 8(28), 2003: 1159–1168.

Alexander, Phinikarides, Kindyni Nitsa, Makrides George, and Georghiou George. "Review of photovoltaic degradation rate methodologies." *Renewable and Sustainable Energy Reviews*, 21, 2014: 143–152.

Arons, A. B., and M. B. Peppard. "Einsten's proposal of the photon concep—A translation of the Annalen der Physik Paper of 1905." *AmericanJournal of Physics*, 5, 1965: 367–374.

Coelho, Roberto Francisco, and Denizar Cruz Martins. "An optimized maximum power point tracking method based on PV surface temperature measurement." In

Sustainable Energy: Recent Studies, by Alemayehu Gebremedhin. IntechOpen, 2014.

Fahrenbruch, A. L., and R. H. Bube. *Fundamentals of Solar Cells*. San Francisco: Academic, 1983.

Fraunhofer Institute for Solar Energy Systems. "Photovoltaic report." 2020.

Green, A. M. *Solar Cells Operating Principles, Technology, and System Applications*. Englewood Cliffs: Prentice-Hall, 1982.

Kibria, Mohammad Tawheed, Akil Ahammed, and Saad Mahmud. "A review: Comparative studies on different generation solar cells technology." *Proceedings of 5th International Conference on Environmental Aspects of Bangladesh*, 2015: 51–53.

Kyocera Installation Manual. "Installation manual for the KC-series of solar photovoltaic power modules." 2009.

Lasnier, F., and T. G. Ang. *Photovoltaic Engineering Handbook*. New York: Adam Hilger, 1990.

Mingsukang, Ammar Muhammad, Mohd Hamdi Buraidah, and Abdul Kariem Arof. "Third-generation-sensitized solar cells." In *Nanostructured Solar Cells*, by Narottam Das, 1–9. IntechOpen, 2017.

Mohammadnoor, Imamzai, Aghaei Mohammadreza, Hanum Md Thayoob Yasmin, and Forouzanfar Mohammadreza. "A review on comparison between traditional silicon solar cells and thin-film CdTe solar cells." *Proceedings National Graduate Conference 2012*, 2012.

Moller, H. J. *Semiconductors for Solar Cells*. Norwood: Artech, 1993.

Osterwald, C. R., S. Rummel, A. Anderberg, and L Ottoson. "Degradation analysis of weathered crystalline-silicon PV modules." *9th IEEE PV Specialists Conference*, 5, 2002: 20–24.

Ruther, Ricardo, and G. Kleiss. "Advantages of thin-film solar modules in façade, sound barrier and roof-mounted PV systems." *Porf. EuroSun*, 1996: 862–867.

Smets, A., K. Jager, O. Isabella, and M. Zeman. *The Physics and Engineering of Photovoltaic Conversion, Technologies and Systems*. Cambridge: UIT, 2016.

United States Department of Energy. *Basic Photovoltaic Principles and Methods*. Colorado: Technical Information Office, 1982.

Villalva, M. G., J. R. Gazoli, and E. R. Filjo. "Comprehensive approach to modeling and simulation of photovoltaic arrays." *IEEE Transactions on Power Eletrocnics*, 5(24), 2009: 1198–1168.

Chapter 7

Functional-Form Sufficiency Achieved by Biased Random Walk

Behavioral Model in Architectural Bioactive Design

Yomna K. Abdallah, Alberto T. Estévez,
Neveen M. Khalil, Diaa El Deen
M. Tantawy, and Mostafa M. Sobhy

INTRODUCTION

Microbial world offers most spatiotemporal coherent forms, especially aligning their formal compositions with their physiological behaviors. One crucial example of these behaviors is the active migration of cells, which is essential for a number of biological processes. In this behavior, the cellular microenvironment heterogeneity which provides the substrate for the cell's migration controls the resulting pattern and complexity of the cellular behavior (Hatzikirou et al., 2010). Two distinct strategies of cells responding to an environmental stimulus are either the cells are following a certain direction and/or the environment imposes only an orientation preference. One example of directed cell motion in a dynamically changing environment is called Chemotaxis which is mediated by diffusible chemotactic signals (Hatzikirou et al., 2010). In a chemotaxis behavior, microorganisms move toward higher concentrations of nutrients (chemoattractants) and away from toxins (chemo-repellents). Chemotaxis by cells offers a mechanism for aggregation. If the cells secrete a chemoattractant, then a random fluctuation in the microenvironment will cause the increase of the local chemoattractant concentration, attracting more cells to the chemoattractant direction, such that these cells in turn again increase the chemoattractant concentration in a positive feedback loop. Eventually, the cells will all move into one or more compact clusters depending on the range of diffusion of the chemoattractant and the response and sensitivity of the cells (Alber et al., 2003).

In order to simulate this intelligent behavior, the dynamic mathematical models analyze and model cellular behavior in a bio-learning process that leads to sufficient biodigital design applications. The design and construction of a biological-based mathematical model demands a critical consideration of the mechanisms that underlie a biological process. The model recapitulates the biological system's behavior and summarizes all of the data that it was constructed to replicate. Moreover, mathematical simulations can be carried out often in seconds with almost no cost. In addition, the biological model behavior can be explored in conditions that could not be achieved in the laboratory, giving the full insight into the cellular behavior of an analyzed biological process as every aspect of the model behavior can be observed at all time-points (Ingalls, 2012). Besides the flexibility in changing states, mathematical models also include parameters that characterize interactions among system components with the environment. A change in the value of a model parameter corresponds to a change in environmental conditions or in the system itself. These values can be varied to explore the system behavior under perturbations or in altered environments (Ingalls, 2012).

One example of chemotactic cellular behavior under different microenvironments heterogeneity is the fungal mycelia. The rules that control the colony are local but lead to patterns on a large scale. This could be represented by employing cellular automata rules to describe mycelia spatial dynamics. Cellular automata models have been proposed for studying the emergence of collective macroscopic behavior emerging from the microscopic interaction of individual components, such as molecules, cells, or organisms. Cellular automata are systems consisting of a large set of basic discrete state elements interacting via a given set of local rules that reveal in a global behavior of the entire system (Alber et al., 2003). Mathematically, a cellular automaton is a tuple (T, M, Φ), where T is a lattice such as integers or higher-dimensional integer grid, M is a set of functions from an integer lattice with one or more dimensions to a finite set, and Φ is a locally defined evolution function. The lattice in M represents the "space" lattice, while the one in T represents the "time" lattice. Cellular automata are often based on a two-dimensional grid where the individual cells change their state discretely according to a uniform and constant set of rules involving the states of the cells and their interaction (Halley et al., 1994). Boswell et al. (2006) built a cellular automaton model for detailed understanding of the influence of various environmental heterogeneities on the events at the hyphal level (Boswell et al., 2006). However, cellular automata simplicity and limitation do not fully represent the complexity of real chemotaxis cellular behavior that is imposed by various parameters including time limits, concentration thresholds, and multi-stimulus environments. Thus, a more realistic mathematical modeling approach is required to model fungal mycelium formation behavior in response and synchronization to the chemotaxis behavior. A biased random walk model is potentially an adequate and simple method based on master equation in which the rate of change of the value of a system variable at a given point is related via transition probability rates, to the values of this variable at a number of neighboring points (Hillen et al., 2000). In this research, biased random walks are employed through bio-learning design procedures to apply the mathematical rules of fungal cells chemotaxis in form finding process. As a step-in coupling form and function of the design of a cluster that incubates the fungal culture of *Aspergillus sydowii* NYKA 510 to produce the laccase enzyme for further industrial usage to achieve a bioactive design (Abdallah et al., 2019).

Dynamic Mathematical Models for Cellular Behavior in Microenvironments

A mathematical model is a physical representation of mathematical concepts, equations, or reality. Physical mathematical models include reproductions of geometric figures made of different substances. Essentially, mathematical models analyze anything in the physical or biological world, whether natural or involving technology if it can be described in terms of mathematical expressions. Thus, optimization and control could be used to model processes, patterns, and other phenomena in biological systems (http://www.mat.univie.ac.at/~neum/papers.html#model).

Mathematical Model Properties

In order to gain full control over a mathematical model, the designer should be aware of the key features of it. In the following section, the main characteristics of a mathematical model are discussed:

- Linearity vs. nonlinearity: A linear dynamic mathematical model imposes linear relationship in all interactions among its components, which is a restrictive condition; consequently, linear models display only a limited range of behaviors. On contrary, nonlinear relations do not follow any specific pattern, resulting in difficulty to address them with any generality. On example of nonlinearities in biochemical reactions is saturations. In this case one variable increases with the another is decreasing (Ingalls, 2012).
- Global vs. local behavior: Nonlinear dynamic models exhibit a wide range of behaviors. The detailed analysis of the global behavior of such models is complicated. Instead, focusing on specific aspects of the system behavior would be more effective and realistic. This is achieved by limiting attention

to the behavior near particular operating points, taking advantage of the fact that over small domains, nonlinearities can always be approximated by linear relationships. This method is called local approximations and is used often in biological modelling as most of self-regulating systems spend much time operating around specific nominal conditions (Ingalls, 2012).

○ Deterministic models vs. stochastic models: A mathematical model is called deterministic if its behavior is exactly reproducible. Although the behavior of a deterministic model is dependent on a specified set of conditions, no other forces have any influence, so that repeated simulations under the same conditions are always identical; unlike stochastic models that allow for randomness in their behavior. The behavior of a stochastic model is influenced by specified conditions and unpredictable forces. Repeated stochastic simulations thus yield distinct samples of system behavior (Ingalls, 2012).

○ Memory versus irreversible decision-making: Adaptive systems are able to eventually ignore, or forget a persistent signal; the opposite behavior enables some systems to remember the effect of a transient signal. This memory effect is achieved by a bistable system. As an input pushes the state from one basin of attraction to the other resulting in a persistent change even after the input is removed. This steady-state response of a bistable system is a switch-like behavior. Either the state is perturbed a little and then relaxes back to its starting point, or it gets pushed into a new basin of attraction and so relaxes to the other steady state (Ingalls, 2012).

Cellular Automata

As mentioned previously in this chapter, cellular automata models study the emergence of collective macroscopic behavior emerging from the microscopic interaction of individuals in biological systems. These cellular automaton models are based on qualitative conclusions of the biology to design simple local evolution rules to examine the characteristics of microbial cellular behaviors in different heterogeneity environments, for the modeling and simulation of these complex systems (Hogeweg, 1988; Men et al., 2013; Hatzikirou et al., 2010). Typical cellular automata consist of discrete agents which occupy some or all sites of a regular lattice. These agents have one or more internal state variables, which may be discrete or continuous, and a set of rules describing the evolution of their state and position. Both the movement and change of state of agents depend on the current state of the particle and those of neighboring particles. Normally the evolution rules apply in steps, for example, a motion step followed by a state change or interaction step that can be synchronous or stochastic (Alber et al., 2003). The main limitations of cellular automata include their lack of biological sophistication in aggregating subcellular behaviors, difficulty of going from qualitative to quantitative simulations, artificial constraints of lattice discretization, and the lack of a simple mechanism for rigid body motion (Alber et al., 2003). All these factors limit the capacity of cellular automata model to simulate natural fungal cells chemotactic behavior that depend on a continuous domain of concentration values of the chemoattractant while reacting only to specific threshold values (Tsompanas et al., 2017). However, theoretically, cellular automata are attempting method of biological cellular behavior modeling due to their completely self-organized large-scale behaviors. An individual cell has no sense of direction or position. It can only respond to signals in its local environment (Men et al., 2013; Alber et al., 2003).

The main features of cellular automata are: finite number of cells (automaton vs. space), uniform (automata or connections.), deterministic, with or without external input/output, static versus dynamic (in static systems the number of cells and interconnections of cells, whether uniform or nonuniform, are fixed; in dynamic systems, cells and their interactions are generated as part of the behavior of a cell) (Hogeweg, 1988). However, synchrony is inconsistent with the localness of cellular automata. To achieve synchrony at the global scale, a global clock is needed (Hogeweg, 1988). The previously described limitations, or the basic cellular automaton, have led to more specialization toward simulation of natural cellular systems. The random walk is a stochastic random process that describes a path that consists of a succession of random steps on some mathematical space. It is mainly used to simulate the path traced by a molecule as it travels in a liquid or a gas (Pearson, 1905; Wirth, 2015; Wirth et al., 2016; Piña et al., 2011). The term most often refers to a special category of Markov chain which is a stochastic model describing a sequence of possible events in which the probability of each event depends only on the state attained in the previous event (Gagniuc, 2017; Serfozo, 2009; Rozanov, 2012). This makes them a potent tool to simulate the biochemical reactions in real physical mediums. Moreover, the possibility of scaling the random walk models to architectural design applications serves the coupling of form and function in a physiological coherency of the bioactive design. However, the absolute random behavior of the random walk model does not resemble the biased behavior of chemotaxis. This leads to a search for controlled randomness, starting from a totally random behavior of exploration and moving slowly to a vectored motion toward the targeted molecule. This behavior is called biased random walk. A biased random walk model is a more specified type of random walk that is a time path process in which an evolving variable jumps from its current state to one of various potential new states. Unlike in a pure random walk, the probabilities of the potential new states are unequal, as there is a source of attraction (a chemoattractant) that controls the motion of the agents (the cells) toward certain vector in the space. This characteristic of biased random walks makes it more unpredictable and stochastic process that could simulate sudden biochemical reactions corresponding to unexpected sudden stimulus, and that make it the most attempting to simulate biochemical processes such as chemotaxis.

Fungal Cells Biased Random Walk Behavior Corresponding to Chemoattractant

Nutrients that are considered the main attraction for any living organism control this organism motion in their medium. In the current case of filamentous fungal hyphae, the nutrients uptake process results in the fungal hyphae formation. These thread transportation tubes are formed based on two different spatiotemporal strategies: the exploration and the exploitation. As mentioned previously, the most essential function of cell motion is substrate uptake. This is achieved by using internal substrate as a fuel for the cells to build more transportation tubes (hyphae) to acquire external substrate by active transport across the plasma membrane. The acquisition rate must therefore depend on the amount of internal substrate available to perform the active transport, the amount of external substrate available for absorption, and the hyphal surface area over which the absorption occurs (Boswell et al., 2006; Fricker et al., 2008). Thus, accordingly, cell motion or translocation is categorized into two different mechanisms that correspond to the exploration and the exploitation strategies mentioned before. These mechanisms are responsible for nutrient reallocation in many fungal strains—the simple diffusion and the active movement of intracellular metabolites from regions of local excess to regions of local scarcity (Boswell et al., 2006). Only newly formed hyphae (and associated hyphal tips) use active translocation, while older, established hyphae use diffusion as the major means of internal nutrient reallocation. As these translocation mechanisms align with the two distinct mycelial formation strategies, exploration and exploitation. The exploration is adopted in low-nutrient environments and features fast-moving hyphal tips coupled with minimal branching, resulting in a sparse mycelial network. Exploitation is adopted in high-nutrient conditions and features slower-moving hyphal tips and increased branching, resulting in a dense mycelial network. The translocation process controls exploration and exploitation predominantly (Boswell et al., 2006). Changes in the active translocation process alone account for the switch between exploration and exploitation strategies (Boswell et al., 2006). Mycelial systems tend to become more open with time as they become larger. Patterns are modified by the quantity and quality of the resource from which the mycelium is extending, along with nutrient status (Zakaria et al., 2002), and microclimate (Fricker et al., 2008).

Mycelial networks are also persistent; they employ a certain "sit and wait" strategy to adjust with the continuous change in substrate availability and concentration in the surrounding environment. There is considerable scope of communication within the mycelial networks, since hyphae maintain continuity with their immediate ancestors and with neighboring regions forming cross-links. This results both radially and tangentially in systems with many connected loops (Fricker et al., 2008). Not only does the mycelium respond by changes to system architecture but also with physiological responses. There is highly coordinated uptake, storage, and redistribution of nutrients throughout the network. Many factors, including the overall nutritional status of the mycelial system, the distribution and quantity of colonized and newly encountered organic resources, and the main sites of uptake, storage and demand for carbon and mineral nutrients affect the process of translocation (Boddy et al., 2006).

In order to mathematically model this switching behavior of exploration and exploitation performed by the fungal hyphae, the biased random walk model is employed. Modeling the fungal mycelia is scale varied as the indeterminate growth habit of fungi incorporates operating over scales ranging from the (sub) micron to the kilometer. Further complexity is added when modeling physical and nutritional heterogeneity of the host environment, and the resultant diverse sensing and response events that media hyphal growth and thus determine network architecture (Davidson et al., 2011).

Modeling nutrients uptake by fungal mycelium has been attempted using various approaches, including partial differential equations, autonomous agents (Hatzikirou et al., 2010), and network-based approach, which achieved a great insight into the analysis of this process. One way to achieve this is to calculate shortest path distances from each node to every other (Fricker et al., 2008). Another way is the "cell" models, which are ideal for modeling the movement of individual particles since each cell (or site) can take a value corresponding to the current state of that cell. Thus, complying with the main memorylessness rule of the Markov process makes it possible to utilize random walks as a tool of simulating this fungal cellular behavior in nutrients uptake process. One good example of employing cell models in fungal hyphae formation, in response for the search and uptake of nutrients, is the study of Boswell et al. (2006) that used a combination of "cell" models for modeling internal/external substrate and hyphal tips formation and "bond" models used for modeling active/inactive hyphal state. However, this cellular approach is not suitable for modeling the development of a network since adjacent cells need not to be connected in a more continuous method (Boswell et al., 2006). The bond-based approach method is similar in concept to biased random walks, as both result in the derivation of a master equation in which the rate of change of the value of a system variable at a given point is related via transition probability rates, to the values of this variable at a number of neighbouring points (Hillen et al., 2000). The discretization procedure of these models allows certain key processes, including hyphal inactivation and reactivation, anastomosis and branching to be treated in a more detailed manner.

In simulation of the chemotaxis behavior that refers to the motion induced by the presence of chemical attractant in the environment, the fungal hyphae moves toward higher concentrations of nutrient chemoattractants and away from toxins and other chemorepellents. When exposed to a gradient of chemoattractant or chemorepellent, the cells bias their random walk by twitching less frequently when moving in the "good" direction (toward the chemoattractant) and vice versa in case of chemorepellent. Once a cell finds itself again in a uniform environment, it returns to the original tumbling frequency of the normal random walk (Alber et al., 2003; Halley et al., 1994).

NYKA 510 Pavilions

The NYKA 510 Pavilions are bioactive lightweight structures aiming to integrate the bioagents in the built environment and introduce them to a public space, in order to augment the biophilic tendency of users to accept microbes as productive and useful design elements in the built environment, achieving double environmental benefits by incubating the fungal cultures of *Aspergillus sydowii* NYKA 510, to produce the laccase enzyme that was proven to play key role in bioelectricity generation in a microbial fuel cell, and to recycle organic agro-wastes by using banana peel as the main carbon source in the growth medium of the fungal culture. All these environmental functions of the proposed pavilion design are combined with its main function of shading an opened public space, providing the experience of growing alive architecture, where the increase of the shade hues is concurrent with the growth process of the fungal cultures. The growth medium with the fungal spores are injected in the free formed transparent tubes that form the pavilion building units, these tubes incubate the fungal culture for 7 days/each cluster, then the exhausted medium with the fungal spores are evacuated from the tubes by simple valves and collected in containers to be further processed for laccase enzyme extraction (Abdallah et al., 2019). The pavilion is suitable in moderate climates where temperature ranges from 25–35°C. In the form-finding process of the pavilion, the previous analysis of the chemotaxis behavior of the fungal cells in nutrients uptake was utilized, employing a biased random walk equation for the form finding process of a pavilion's design. The equation is based on the cellular behavior of the filamentous fungi *A. sydowii* NYKA 510, of the chemotaxis behavior toward carbon source molecules concentration in the liquid medium space. The equation is based on statistical biological data obtained in a previous work conducted by the authors, where an experimentation was conducted to optimize the growth medium constituents and their concentrations (Abdallah et al., 2019).

In the design process, the master equation of biased random walk is based on capturing spate-time status of each agent (fungal cell) in order to record the behavior of the fungal cells in the liquid media space (the microenvironment) in their chemotaxis behavior as explained before. Those agents (fungal cells) are modifiable in the digital mathematical platform, in terms of count, vector of biased walk toward the target molecules of (carbon source) resulting in a chemoattractant effect. In this current design study, the fungal cells count was defined to a limited number in order to gain clear insight of the walk path and avoid tangled web of criss-crossed biased random walks of a huge count of cells.

Design Methodology

For designing the mathematical model, the parametric software Rhinoceros 3D + Grasshopper (Boids plugin) was employed to simulate the cellular behavior of *A. sydowii* NYKA 510 in the chemotaxis processes. Boids plugin in Grasshopper provide a wide range of tools simulating and modeling microbial behaviors, using various mathematical models and their incorporated forces, including among others, the agent-based models, and different types and classes of cellular automata. In this research and according to experimental procedures conducted and statistical results obtained, simulating the nutrients search processes requires defining the main controlling constraint, which is the carbon source concentration, the banana peel at 15.1 g/l.

Forming Master Equation of the Biased Random Walks

Motion type: Initially, fungal cells exhibit 3D spatial navigation in searching the liquid medium for the carbon source molecules. As soon as these cells begin to sense the optimum concentration of the carbon source, a chemoattractant is released by the cells and their neighbor cells and force them to bias their motion. Thus, their motion defines two spatial domains: polar domain (around 360° surrounding the spatial domain around the fungal cell), which resembles the initial random search in the liquid medium, and the vectored domain, under the chemo-physical attraction caused by the carbon source molecules that starts to bias the cells searching to a directed and vectored motion.

Constraint (chemoattractant): Banana peel concentration as the carbon source in growth media was defined to fifteen approximating statistical result of 15.1 g/l. This constraint determines the direction of fungal cell biased random walk, which defines the vector of motion of carbon source molecules inside the liquid media space-defining carbon source molecules specific position in medium space.

Parameters

Agents/walkers: These represent the population of fungal cells inside the medium. This parameter could be adjusted according to two objectives. The first, resembling the full population of the inoculum in medium to obtain a picture of the whole colony behavioral pattern; this implies running simulation by varying the spatial position of the carbon source molecule while recording history, in order to animate the searching motion of the whole population in time. This method will result in a large amount of computational data that won't be easy to handle in further design steps and will only give an approximate analysis of the real behavior, as it ignores the fact that fungal cells diffuse in media which implies the distribution of fungal population in medium space. The second option is the regional analysis of fungal cells distribution in medium, which is a more realistic analysis as it simulates different population numbers randomly complying with the fact that fungal cells are not statically located in one spot in the medium, but are in continuous dynamism across different regions of it. This implies that fungal cells first move randomly in different regions of the medium. Thus, the authors chose to randomly change the agents count with consideration to computational capacity in this study.

Chemotaxis and searching spatial domain: These parameters are defining the searching movement of fungal cells in response to diffusion balanced with the attraction of chemoattractant reaction caused by carbon source molecules' concentration. This sums the resultant transition of fungal cells from two contradictory forces: the diffusion behavior that causes the agents to wander freely in the medium space in three dimensions around 360° at each point in space, and the attraction force toward carbon source molecules that they will bind with (vectored motion). This parameter was set to 360° as it is the maximum search domain at each point in space using the spherical spatial grid.

Random initial sits of the chemical reaction: This parameter is to propose different random points inside the medium that the oxidation reaction starts in. This parameter is random, considering that in an almost homogenous medium, the concentration of nutrients is almost equal at each point of the space resulting in an equal probability of distribution of active sites of the oxidation reaction. However, the simulated behavior captures the first-time frames of the reaction, and is to adjust the cellular biased random walks simulation to specific spatial coordinates.

In this design, the author proposed further parametrization of the code by presenting more iterations that are achieved by moving the initial site of reaction along a defined curve, implied by the designer, and that could vary according to different design physical or spatial layouts. This curve is divided to a number of points that could be increased. By running the simulation, the number slider changes which point along the curve is selected, thus the agents head toward it. The code is shown in figure 7.1.[1]

In figure 7.1, the "cells" components define points on referenced curves in Rhinoceros 3D. Each "cell" component is evaluated to choose a point on the referenced curve to be the site of the reaction, these reaction sites' points are defined by the optimum concentration of carbon source extracted from experimental statistical data (Abdallah et al., 2019). This main simulation equation exhibited in figure 7.1 is a random walker loop; start and end components set the simulation of the defined population of fungal cells (10 cells) in searching the domain of 360° around each of these cells. There are serving components that are mandatory input data for these two main loop components. These are the "Populate 3D" and the "Random Vector," to ensures the random motion of fungal cells' walks inside the domain surrounding each of them in the liquid medium. To bias this random walk loop, the "Adhere" component was added to the loop to achieve an attraction force of the fungal cells toward the optimum concentration of carbon source simulating the chemotaxis process. This adherence was defined to the point of reaction site (on the referenced curves from Rhinoceros 3D) and fixed with a ratio to prevent total adherence that eliminate the random walk loop effect. To achieve this, a multiplication and a summing component were added to alleviate the adherence effect to achieve the biased random walk. Figure 7.2 shows the form similarity between a scanning electron microscopy image of the fungal culture of 7 days old, *Aspergillus sydowii* NYKA 510 and the resulting behavioral pattern from the biased random walk equation.

Bioactive Design Criteria

The pavilion preforms outdoor shading and incubates the filamentous fungal strain *A. sydowii* NYKA 510 to produce laccase enzyme to be further employed in electricity generation in microbial fuel cells and other industrial applications (Abdallah et al., 2019).

The formal design maximizes the bioreceptive surface of the sophisticated form of the biased random walk for microbial colonization. Areal propagation in transparent tubes achieves maximum visual engagement in the space, and integration of bioactive agents in the architectural design.

Spatial propagation of spatiotemporal growth in bioactive design is achieved by clustering or pixilation. The functional clusters, which take the form of knots, are confined, controllable, and easy to manipulate in space. More clusters could be added or removed according to the design spatial expansion case for needed shading, and could be arranged along different axes in space. To achieve bioactive formal criteria, the following aspects are considered: time, accumulated frames, and phases. These are attained by the cumulative effect of biased random walker equation that rules the cellular behaviour of the bioactive agents (fungal cells). The algorithm was designed in a record history mode to draw the bioactive agent track through the whole simulation process in a nonstop line path. This path is the final design form per cell as exhibited in figure 7.3.

Exhibiting time lapse criterion also, the growth phases with its distinguished features of the *A. sydowii* NYKA 510 fungal culture added a differential spatiotemporal real time effect, as the texture and color of the newly inoculated fresh media with dispersed spores in light yellow liquid turns into light green and bluish green and the distinguished texture of the hyphae to the final dense textured culture as exhibited in figure 7.4.

For designing time frames in this study, scale was considered according to its contribution to the final formal design ratio. The effect of changing colors and textures was considered as hue in the form of pixels to create a smooth, blended and integrated effect as shown in figure 7.3 and figure 7.4. It was also considered in the mass customization processes, as these phases exhibit different growth forms across different clusters, as one cluster would be in mature phase, and the other is at the initial phase. Thus different clusters will always exhibit different colours and textures of growth according to their growth state as shown in figure 7.4.

The random cellular path that forms the cluster provides infinite iterations of mass customizations, and final forms. The intersection and integrity of the opened transparent clusters create various spatial relations between different parts of different clusters emphasized by the different colors and textures of the different growth phases resulting in a new visual conception from different views around the 360° in space and different axes as shown in figure 7.3 and figure 7.5.

The fluidity of the mass is achieved in each cluster through the chaotic random form resulting from the cellular path of the biased random walk on the level of the single cluster. In the other hand, pixilation is achieved by the ability to repeat these cluster infinite times in space along different axis, in different orientations and relations. This defines the design as both fluid and pixeled through manipulating scale, mass customization and spatial propagation criteria. The resulting mass is an unpredicted fluid pixeled mass that has chaotic free edges and undetermined solids and voids to exhibit integration in all aspects and create continuity in visual conception.

The spatial propagation dealt with arranging clusters in space and arranging their relations according to visual conception in different views around 360°. This was predefined by the algorithm of biased random walk for multiple cells. It defined the margins between different clusters and their location in space. Spatial orientation was limited by the horizontal configuration of the cluster. However, the design cluster is free in orientation (rotation) as it is a free fluid form that could be defined in any position. The cluster pro-liferation axes (axial growth) were defined as aerial, longitudinal, and orbital axes. The proliferation method realized in this study was branching as it is the dominant proliferation method of the bioactive agent *A. sydowii* NYKA 510 hyphae formation. This branching method inspired the biased random walk algorithm. Thus, achieving coupling between form and function as well as spatial propagation through branching along aerial-longitudinal/orbital axis as shown in figure 7.3 and figure 7.5.

CONCLUSION

The current study proposed a biased random walk model extracted from fungal behaviour of nutrient search and chemotaxis processes. This biased random walk model was based on statistical data obtained from previous experimental work conducted by the authors to optimize growth conditions of fungal strain *A. sydowii* NYKA 510 to the production of laccase enzyme. The current work designed the biased random walk equation with the use of algorithmic design software Rhinoceros 3D + Grasshopper (Boids plugin). The resultant equation was utilized in the design of a pavilion that couples form, function of a bioactive design through applying the behavioral pattern extracted from the fungal cells in the form-finding process of the pavilion. The pavilion form is composed of clusters of transparent tubes that will incubate the fungal strain to produce the laccase enzyme that will be har-nessed from the pavilion to be utilized in the production of bioelectricity in microbial fuel cells. The authors further analyzed this approach of coupling form and function to propose bioactive design criteria that are applied in the current design.

CONFLICTS OF INTEREST

Authors declare there is no conflict of interest. This research was conducted as part of PhD programme granted by the Egyptian Government, Ministry of Higher Education, Cultural Affairs and Missions Sector, 2019/2020.

NOTES

1. Figures 7.1 through 7.5 available online at https://rowman.com/ISBN/9781793625021/Sustainable-Engineering-for-Life-Tomorrow.

REFERENCES

Abdallah, Y. K., Estévez, A. T., Tantawy, D. M. A., Ibraheem, A. M. M., Khalil, N. M., "Employing Laccase producing *Aspergillus sydowii* NYKA 510 for cathodic biocatalyst in self-sufficient lighting microbial fuel cell", *Journal of Microbiology and Biotechnology*, vol. 29, num. 12, pp. 1861–1872, 2019.

Alber, M. S., Kiskowski, M. A., Glazier, J. A., Jiang, Y., Rosenthal, J., Gilliam, D. S., "On cellular automaton approaches to modeling biological cells-mathematical systems theory in biology", *Communications, Computation, and Finance*, vol. 1, num. 39, 123–139, 2003.

Boddy, L., Jones, T. H., "Mycelial responses in heterogeneous environments: Parallels with macroorganisms", In *Fungi and the Environment*, Cambridge University Press, Cambridge, UK, pp. 112–140, 2006.

Boswell, G. P., Jacobs, H., Ritzc, K., Gadd, G. M., Davidson, F. A., "The development of fungal networks in complex environments", *Society for Mathematical Biology, Bulletin of Mathematical Biology*, vol. 69, pp. 605–634, 2006.

Davidson, F. A., Boswell, G. P., Fischer, M. W. F., Heaton, L., Hofstadler, D., Roper, M., "Mathematical modelling of fungal growth and function", *IMA Fungus*, vol. 2, num. 1, pp. 33–37, 2011.

Fricker, M., Bebber, D., Boddy, L., *Chapter 1 Mycelial Networks: Structure and Dynamics*, British Mycological Society Symposia Series, 2008.

Gagniuc, P. A., *Markov Chains: From Theory to Implementation and Experimentation*, John Wiley & Sons, NJ, pp. 1–235, 2017.

Halley, J. M., Comins, H. N., Lawton, J. H., Hassell, M. P., "Competition, succession and pattern in fungal communities: Towards a cellular automaton model", *Oikos*, vol. 70, num. 3 (Sep., 1994), pp. 435–442, 1994.

Hatzikirou, H., Deutsch, A., "Lattice-gas cellular automaton modeling of emergent behavior in interacting cell populations", *Understanding Complex Systems*, pp. 301–331 2010, Springer, Berlin, Germany.

Hillen, T., Othmer, H. G., "The diffusion limit of transport equations derived from velocity-jump processes", *SIAM Journal of Applied Mathematics*, vol. 61, num. 3, pp. 751–775, 2000.

Hogeweg, P., "Cellular automata as a paradigm for ecological modeling", *Applied Mathematics and Computation*, num. 27, pp. 81–100, 1988.

Ingalls, B., *Mathematical Modelling in Systems Biology: An Introduction*, University of Waterloo, 2012, Waterloo, Canada.

Men, H., Zhao, X., "Microbial growth modeling and simulation based on cellular automata", *Research Journal of Applied Sciences, Engineering and Technology*, vol. 6, num. 11, pp. 2061–2066, 2013.

Pearson, K., "The problem of the random walk", *Nature*, vol. 72, p. 294, 1905.

Piña, C. A., García, D. G., Huosheng, H., "A composite random walk for facing environmental uncertainty and reduced perceptual capabilities", *ICIRA*, 2011.

Rozanov, Y. A., *Markov Random Fields*, Springer Science & Business Media, p. 58, 2012.

Serfozo, R., *Basics of Applied Stochastic Processes*, Springer Science & Business Media, p. 2, 2009.

Tsompanas, M. A., Adamatzky, A., Sirakoulis, G. C., Greenman, J., Ieropoulos, I., "Towards implementation of cellular automata in microbial fuel cells", *PLoS One*, vol. 12, num. 5, p. e0177528, 2017.

Wirth, E., *Pi from Agent Border Crossings by Net Logo Package*, Wolfram Library Archive, 2015, Champaign, USA.

Wirth, E., Szabó, G., Czinkóczky, A., "Measure landscape diversity with logical scout agents", *ISPRS International Archives of the Photogrammetry: Remote Sensing and Spatial Information Sciences*, vol. XLI-B2, pp. 491–495, 2016.

Zakaria, A. J., Boddy, L., "Mycelial foraging by Resinicium bicolor: Interactive effects of resource quantity, quality and soil composition", *FEMS Microbiology Ecology*, vol. 40, num. 2, pp. 135–142, 2002. http://www.mat.univie.ac.at/~neum /papers.html#model

Chapter 8

Circularity in Kitchen Design and Production Business

A Sustainable and Disruptive Model

Julia Ann Baker, Marco Samuel Moesby Tinggaard, Peter Enevoldsen, and George Xydis

INTRODUCTION

Climate change is perceived to be a global problem and is on many agendas around the world (Solomon, 2008). As one of its key priorities, the European Union (EU) has waste management on several programs, based on the following waste hierarchy: prevent, reuse, recycle, energy recovery, and the least preferred disposal (European Union Council, 1999). This hierarchy's vision is to reroute waste from landfills to energy recovery (Lausselet et al., 2017; Xydis et al., 2013) preventing the loss of non-sustainable resources. Consumption has for long been pondered as a sign of wealth (European Union, 2015). The current linearity of product ontogeny needs to be questioned toward a more efficient usage of resources (Ritzén and Sandström, 2017; Nikas et al., 2018).

The strict definition of the circular economy (CE) is the obvious way to reject linearity. Ritzén and Sandström (2017) simply define CE as the way to avoid disposal by closing the material loop within the product lifecycle in order to decrease the use of resources and subsequently the demand for energy. Unlike conventional recycling, CE emphasizes to the reuse, remanufacture, refurbish, and repair the items, parts, and materials (Korhonen et al., 2017) aiming at maintaining them for a longer time in the market (Amui et al., 2017), while also encouraging sustainable energy, when and where possible (Aviles et al., 2020).

Manufacturing

Manufacturing is the process that converts raw material into finished goods meeting the expectation or specifications of the receiving customer (Koscis and Xydis, 2019). This is often done by a man and machine setup, which is then scaled to the desired size for the manufacturer. Large companies have multiple divisions of manufacturing setups (Business Dictionary, 2017). The manufacturing paradigm goes back to centuries and is vital for a nation's economy. The vitality can be attributed to the creation of the high-paying jobs the manufacturing industry creates, thus creating lasting wealth for the citizens of the nation (Hu, 2013). Manufacturing has undergone paradigm changes since its dawn. Until now, there have been four paradigms within manufacturing. The first paradigm is the craftsman paradigm. The craftsman manufactures the product to the customers' direct wishes at a high price and low speed (Hu, 2013). The second paradigm is mass production. Because of moving assembly lines it became possible to manufacture goods at a high speed and at a lower cost. The United States was first at implementing this new way of manufacturing (Duguay et al., 1997). The implementation of the moving assembly line by Henry Ford is considered a cornerstone in the implementation of mass production (Hu, 2013) letting the production line set the pace for the workers (Duguay et al., 1997). The two presented paradigms represent the choice the manufacturer needs to make; either customization or mass production in focus (Duray, 2002). The third paradigm is the lean era: the production is optimized with the aim of minimizing waste along with the production process while maximizing value for the customer. The lean paradigm is often associated with the Japanese automotive industry (Hu, 2013). The fourth and newest paradigm is known as mass customization. The goal of mass customization, as a strategy, can be defined as the use of flexible structures in the organization and the processes to enable manufacturing of customized products, which satisfies the customer at the same low cost as would be the case if the product was produced in a mass production setting (Hart, 1995; Pallant et al., 2020). The four paradigms are shown in figure 8.1.[1]

The kitchen manufacturing industry has changed significantly over the last half century, from a non-existing industry to a dominating and decisive objective, within the building and interior design industry. Customization—adding modules together according to the needs of end users (Duray, 2002)—nowadays plays an important role compared to the mass production approach few decades ago. Studies show that mass customization may be finished with the help of putting modules together in associate order that matches the user's requirements (Duray, 2002).

Circular Economy and Manufacturing

Politics that were more liberal ushered the industrial revolution making use of fossil fuels in the manufacturing world concurrently introducing new mechanical innovations (Mathews, 2011). Nonetheless, it also brought a "counter movement" as what Polanyi (2014) called it, which among others oversaw possible negative impacts on the environment. Hence, it is absolutely necessary that economy and sustainability and sustainable resources grow hand in hand (García-Olivares and Solé, 2015). In modern society, it is more and more seen that the ownership model is challenged. Goods are more often discussed as offered services, where the end user does not own the product, but preferably renting the service, leasing the time used. The product belongs to the industry which undertakes the costs of assembling, operating, and maintaining it on a longer horizon, most probably until technological innovations dictate another company policy (Hawken et al., 2000).

Based on this thinking, product service systems (PSS) were environmental research-driven policies which argue on the matter stating that an unchanged path compared to today's practices will lead mathematically humanity to disaster (Tukker and Tischner, 2006). PSS could pave the ground for large enterprises and small- and medium-sized enterprises (SMEs) to create various value propositions—thereby increasing their revenues, while minimizing the consumption of resources at the same time (Avgoustaki and Xydis, 2020; Koroneos et al., 2017). One of the most known nonlinear business models, within the product lifecycle, is CE (Ritzén and Sandström, 2017). This approach unavoidably improved the manufacturers' approach toward materials and sustainability. Under this scheme, companies own the product and, therefore, they want it to last for more. Prolonging the lifetime of the product incentivises decisions for appropriate material usage (Tukker, 2015).

On the other hand, mass customization resulted that kitchen companies—as all furniture manufacturers—are struggling to keep their market share, since end users became more demanding since more parameters have been added into the "game" such as social and environmental demands, sustainability, and the company's marketing strategy (Stirn et al., 2016; Bennett and Graedel, 2000; Intlekofer et al., 2010).

It is true that leasing as a business model has been found in a number of industries but not yet in the furniture one. This study aims at analyzing a perspective as such using the well-known Danish company Kvik A/S, as a case study.

RESEARCH DESIGN

The heart of this study is to acknowledge how CE and production are theoretically linked. The goal is to understand how CE can participate in all the production processes in the near future. The specific study included semistructured interviews (using the snowball method) with experts from the case company (Saunders et al., 2015).

The layout of the study consists of two desk research performed in parallel. The studies are performed by searching for literature in known databases. The literature found is then analyzed using qualitative reasoning to find consensus within the different articles. The knowledge is then condensed into short narratives about the two areas; this is done to build up an understanding of the different realms.

The second part of the study is combining the two theory realms using literature. This is done to understand the link existing between them within the literature of manufacturing and what drivers are present to further enhance the reaction between the areas of theory, thus making sure that this research contributes to the knowledge base regarding grounding the two theory realms onto mutual ground. To tailor this research toward the kitchen industry a search for the two theories within kitchen manufacturing is conducted and described in the last section of part two. The gaps found act as points of interest for the third part.

The third part consists of a literature study and a case study. The desk researched literature study is done on the topics of non-correlating gaps uncovered in the combination done in the second part. While the case study is a deductive testing of the theory found in the literature including a calculation of the potential amount of product saved by the method. Therefore, the two are done partially sequential (figure 8.1). Via a suggestion on how a specific business model can be linked to circularity and CE, can be the starting point on how to apply such a model to other sectors (Rasmussen et al., 2020).

Circular Economy: From Theory to Practice

It is important for companies worldwide to rethink their recycling and reuse strategies (Singh and Ordoñez, 2016). We know that products cannot be recycled forever, since some materials degrade and are less efficient if reused (Haanstra et al., 2017), and one the other hand the Earth cannot digest everything that is going to the landfills. As of 2017, only 11% of global materials are collected and categorized in order to be recovered and a share of almost 20% ends up to the incinerator (Singh and Ordoñez, 2016).

However, the most important is not to change the recycling and reuse system, but alter the mindset of the production companies that they are not any more product sales companies, but more long-term service providers. This new business model is required to prove to the business owners its viability in order to start discussing this mindset change (Sousa-Zomer et al., 2017). However, this move should be considered as the natural evolvement of the one-place selling company to the stakeholders' new business plan.

Surprisingly, there are cases that the market is more ready than believed. Specific scientific works indicate that for certain product demands the customers are willing to trade in the product (or product value) for a service (Nanaki et al., 2020; Haanstra et al., 2017) which has been successfully implemented for long in the automotive industry.

Leasing

Leasing has been around for many years, however mostly known from the automotive industry. Automobile leasing, as a consumer strategy, has increased in popularity; in 1990 approximately 8% of all American cars were leased while at the end of the century that number had increased to around 33% (Trocchia and Beatty, 2003). Households usually choose to lease due to a lower initial cost. Leasing an automobile rather than purchasing and financing can lower both the down payments and the monthly cost because the leasing expenditure only covers the vehicle's depreciation over the term of the lease rather than the total car (Aizcorbe and Starr-McCluer, 1997). In general, leasing and financing make consumption possible in situations that might not be possible otherwise. According to findings made by Trocchia and Beatty, there are four major motives to leasing versus buying decisions in the automotive industry. It is found that consumers opting to lease will possess a higher desire for gratification due to the fact that they can achieve more with less when leasing (Trocchia and Beatty, 2003) which also helps project a more favorable image of themselves.

Households choosing to finance or lease their kitchen spend more money/choose an upgraded model than what they would have done otherwise. Seen from a company perspective trade-ins and leasing help achieve two basic goals (Li and Xu, 2015):

A) Manufacturers influence and can participate in the second-hand market. So far the kitchen producers were absent from this market, with significant market losses.
B) Overall, make end users aware of the new opportunities, trade-in and lease products and services being at the same time environmental friendly (Li and Xu, 2015) as in the car industry (Nanaki et al., 2015).

This work will present the business model gathering information from the above mentioned making things more clear in relation to multiple revenue streams and end users demand (Ausrød et al., 2017).

Business Model Containing CE in Kitchen Manufacturing

To be able to make a new business model that can handle the challenges connected with CE it is important to take the following key points into consideration:

- Getting the organization ready for CE
- Recycling loop for materials
- Design for x
- PSS
- Persuading the customer

Right Organization for CE

CE in the organization means that the whole company thinks within the realm of circular reuse of material. This means that the whole supply chain needs to be tailored to fit the theory. Therefore, it is necessary that a complete loop is created so all activities limiting the possible reuse of material is reinvented toward a more sustainable method. To enable this kind of feedback loops within the company requires an organizational structure where all employees feel confident in contributing knowledge to improving the process.

Recycling Loop

Getting the materials back to the company is an important task, which can be divided into two categories: (1) how to incorporate the old products into the new supply chain as raw material and (2) how to reclaim the products from after the new business model. This can be done by service technicians visiting customers to replace parts.

Design for X

The theory of design for x needs to be incorporated in order to create a well-functioning loop. New parts must incorporate the old interfaces used in prior produced parts in order to ensure the reusing of discarded products. If the compatibility is not incorporated the recycle loop is broken and the CE is not complete.

PSS for Kitchens

The product functions of a normal kitchen shall be replaced with a service that can satisfy the customer. In order for successful PSS, the customer should not care if he buys a service solution or a physical product from the company. For this to work, the PSS shall contain all the same aspects of buying a physical kitchen. Customers in general, are more demanding when it comes to buying products earlier, which influences the PSS, due to the fact that customers need to perceive that they gain more from buying a PSS kitchen rather than a physical kitchen.

Persuading the Customer

From the automotive industry, it is known that customers are likely to favor the cheapest solution with the highest perceived value—most of the times not even related to the environment and its impact (Nanaki et al., 2014). Meaning that the customers get a feeling of higher value from the service solution compared to the price they pay. In addition, the service solution creates the feeling of being an easier task to undertake for the customer.

Financial

The overall financial differences in the new BM can be divided into two categories, the first of them being the knowledge about the customers tending to buy larger/more when they lease since the initial cost is lower. In addition, the monthly cost of ownership is lower due to the customers only paying for the depreciation over the term of the lease. Second, the cost structure of the company needs to incorporate that the products are not sold therefore the company still owns the raw material, which entails a rise in capital binding within the company.

In the last couple of sections, the BM parts have been presented. In the next section, the theory was converted into a BM. Starting with is a short introduction to the case company. This company have a functioning as is BM this was used as a steppingstone to tailor a new BM using the theory.

Kvik A/S

Kvik A/S is a kitchen manufacturer with 140+ stores around Europe and Asia. The headquarter is situated in Vildbjerg, Denmark, which is also their production site. In Vildbjerg, they produce most of the wood products used in their kitchens. They import cabinet doors and fronts from Italy through four external suppliers. "Accessories" such as handles, cutlery trays etc. are imported from China. All imported products are shipped to Vildbjerg wherefrom all orders are shipped to local stores.

Kvik is very customer oriented—they promote themselves on being Danish design for an affordable cost and spend many resources ensuring customer acquisition through marketing and new product launches. Their slogan is "everyone has the right to a cool kitchen," embedding both the cost and design strategy.

The research group has combined the reviewed theory with Kviks business and created a new BM for Kvik in order to visualize the possibilities of incorporating CE, see figure 8.2.

From the business model shown above it can be seen that implementing CE can to some level coexists with the current way of doing business. This means that the implementation can happen in small incremental steps that will enable the organization to cope with the changes. One example could be the recycle loop that with incremental implementation can be scaled overtime to cope with the recycling of materials instead of having to cope with the total amount of the whole business from the beginning. The business model is partially going to incorporate the CE, as seen in figure 8.3, from the perspective of Kvik A/S.

RECYCLE LOOP

Step 1

A potential customer enters a Kvik store and a relationship starts with a salesperson in the store. The initial thoughts of the customer are presented and the salesperson reacts to the information first by showing the customer different showrooms and next by sitting down with the customer and drawing a kitchen. In the drawing phase, the customer chooses which products they want in their kitchen.

The Trade-In Solution

The trade-in offer is designed to make customers switch from their normal preferred kitchen provider to Kvik by offering them a base price for their old kitchen, also kitchens that do not fit the Kvik product platform. In a case where the product platforms match, the cabinets can be used as a part of the new service solution.

When the cabinets do not match the product platform, they are sold off to a key partner. From there they are either sold as used kitchen or dismantled to recover the raw material that then can be fed back into the supply chain. This strategy is also known in the automotive industry as a powerful persuader to get the customer to commit to buying.

Step 2

The customer is introduced to different methods of payment: cash, financing, or leasing. The payment method does not dictate the customers' willingness to purchase, however it does influence the level of gratification the customer seeks. Being able to finance or lease a product heightens the ability for Kvik to upsell and thereby reach a higher gratification of the customer. If another financial model than leasing is chosen the CE cycle ends at this stage.

Step 3

Step 3 is the installation of the PSS. This means the customer receives the service agreed upon by the customer and Kvik. The service includes the kitchen modules, alliances, and extras that are withheld in the lease agreement installed in the house by installers working for the Kvik store.

Step 4

During the leasing period, the customer can call for service, if something is not satisfactory to the customer. An assessment will be made to figure out whether the error is caused by the product malfunction or the customer not treating the product properly.

Step 5

If the maintenance team finds broken parts these are changed in order to prolong the lifetime of the kitchen.

Step 6

When reaching the end of the leasing period a meeting is set up with the customer and the Kvik representative. The focus of the meeting is to figure out if the lease can be prolonged for a new period and perhaps which changes need to be made in order for it to happen. The contract is prolonged and the loop starts over again.

Step 7

When the lease ends or the contract is terminated, the kitchen parts are collected. This process is completed by the installers from the Kvik store. They ensure the parts are disassembled in a manner that makes it possible to reuse the parts. All parts are shipped to the factory in Vildbjerg.

Step 8

The parts from the dismantling process, which directly can be used again, are stockpiled alongside new goods and the parts that are outdated or decomposed is recycled in the most suitable manner to get the most out of the raw material.

Kvik Presentation

The BM was presented to the managerial team at Kvik and in general, they found the strategy interesting. The managerial team had previously discussed the necessity of CE as a company strategy and discussed leasing as a possibility in their sales strategy—however, they had not connected these two ideas into one. Especially the trade-in system was new to them. They knew of the strategy from other industries but had not thought it could be relevant for their type of business. Especially since all kitchen producers have their own individual platform. As an example, Kvik works with XL and XXL cabinets which most of the other retailers do not. So Kvik posed the question: "What do we do if someone wants to buy our product using the trade-in system, but currently has a different retailers' kitchen?" In this situation, we would see the possibility for Kvik to establish a new customer and thereby still encourage them to trade-in the "off brand" kitchen. Thereby creating an excellent customer relationship and by implementing Kviks own platform other retailers will have a difficult time "stealing" customers back—especially if Kvik nurtures their customer relationships.

Another question posed by the managers was what our expectations were for the competitors on the market. Would they follow in the same footprints or would they expect the leasing not to catch on? According to (Danmarks statistik, 2016) the financial leasing agreements across Denmark rose, for the second year in a row, with 7.5% and contributing with a value of 42.3 billion DKK of the total credit amount approved by financing companies in Denmark. Leasing is a growing form of transaction and the trend does not seem to change now.

Kvik has a unique opportunity to become first movers in the kitchen leasing market, and significantly increase their market share if they can create a customer relationship worth changing brands for.

Standard Kitchen Example

The knowledge gained through previous sections gives an idea of how the theory of leasing could work in an industry as Kviks. However, there is a need for understanding the tangible scale of what such a BM would bring to the business in the form of monetary flow, but also what it would mean to the entire world in a CE aspect.

Through interviews with the Operations Manager and one of the Product developers, each kitchen item was divided into three categories, reusable (green), refurbishment (yellow), and scrap (red) depending on the expected level of degradation after a ten-year installation. Green represents the theoretical possibility of using 100% of the materials in each relevant category. Kvik evaluated the reusable percentage to be 100%, but in reality, some kitchens are more worn down than others, so in further research the material usage is calculated to be 90%. The refurbishment is coded in yellow and is set to a reuse percentage of 50. The red category, scrap, is set to a reuse level of 0%.

To evaluate the possible saved amount of product, a standard kitchen is selected by the product development team. The kitchen chosen, seen in figure 8.4, reflects the typical combination of cabinets and items chosen by a customer when purchasing a new kitchen. Kvik estimates that they sell 36,000 kitchens a year.

Standard Kitchen

The standard kitchen is depicted in figure 8.4. In the next section, we combined the knowledge of the standard kitchen and the leasing system to define the potential savings in product amount and CO_2 emissions.

CALCULATION

In order to calculate the savings, some assumptions have been made. First, it has been chosen to rely on Kvik's estimate regarding annual market growth. The company uses 10% annual growth as their guideline. Second, from the standard kitchen example, it has been stated that the green parts and the yellow parts can be reused. Therefore, for this calculation the first estimates of green equal 90% reusability and yellow 50% reusability is utilized.

The savings are calculated over a ten-year period. This has been done using the following method. Year zero is the first year where the take-back system is in order. From there it is estimated that an additional three years is required before the system is ready to incorporate the used parts as part of its production/service. Therefore, the first three years the company is only stockpiling reusable parts and not reselling them. From the third year onwards, it is estimated that 40% of the kitchens sold will contain parts that are reused.

The same 60/40% is used for the rest of a ten-year period.

From the graph in figure 8.5, it can be seen that the total number of new products produced by and for Kvik is reduced from 87,106,826 pieces to 70,317,270 pieces over a ten-year period. In total a reduction of 19 percent.

Kvik on average receives ten trucks with raw particleboard and three trucks with cabinet fronts each week. The latter comes from northern Italy, while the particleboard trucks are shipped from all over Europe with variations each week. In order to create a viable estimate regarding the climate impact through less transport by utilizing the CE strategy, an average of 1,600 km per cabinet front freight from Italy and an average of 500 km per freight particle board freight is selected (Frausing, 2018; Pedersen, 2018). Each truckload weighs between 23 and 28 tons (Frausing, 2018).

According to The European Chemical Industry Council (CEFIC, 2011) the CO_2 impact is calculated by:

• 50g CO_2/ton-km for truck freight.

Front
The impact per shipment:
$1,600km*25.5t = 40,800tkm$
$50*40,800= 2,040,000g\ CO_2$
Which totals to 6,120,000g CO_2 per week.
By incorporating leasing as a BM and thereby only using 60 % new product the emissions per week could fall to 3,672,000g CO_2 per week.

Particleboard
The impact per shipment:
$500km*25.5t=12,750tkm$
$50*12,750=637,500gCO_2$

Total of 6,375,000 g CO_2 per week, which could be reduced to 3,825,000 g CO_2 per week.

Combining the truck freight, the CO_2 emissions reach 12,495,000 g CO_2, also equivalent to 12,495 kg CO_2 per week if continuing the same production strategy as of now. If Kvik decides to introduce CE, they can reduce their CO_2 emissions to 7,497 kg per week. In figure 8.6, the impact of CO_2 over the next 10 years is shown. The first three years there is no reduction due to implementation of the CE strategy.

Over a ten-year period Kvik will currently create 6,497,400 kg CO_2 emissions through trucking freight (only inbound), by using the CE strategy mentioned in this article they can reduce that impact to 3,898,440 kg CO_2 which amounts to a reduction of 40%.

DISCUSSION AND SUGGESTIONS
FOR FURTHER RESEARCH

CE is not new as a concept. Material reuse has always been one of the most important drivers of manufacturers. Moving to mass customisation has increased demand in production, which requires the skills gained from the mass production era. CE is supporting recycling instead of energy recovering (via disposal). In order to go there, increased recycling and reusability is required and such mentality will be supported only if overconsumption is left behind and understand the different dimensions of the new business model. It is important to understand that regarding the CO_2 calculation, only the inbound freight was incorporated into the calculation. The complete emissions calculation should include outbound logistics, which might give a completely different result.

The unanswered questions within the business model might have an impact on the use of the BM by other companies. However, the answers might not even be useful since the companies in the industry are not bound on the same principles.

Developing an implementation plan for Kvik's new business approach or another similar company shall be among the first priorities in the future, since such an implementation has not been done before in kitchen manufacturing, and unforeseen challenges should be addressed in advance.

Digging further, research will prove the financial viability of such an approach not only in the kitchen market, but also in the other manufacturing sectors. Similar business models could be proposed for instance in the wind industry. For example, the wind turbine manufacturer could offer a long-term lease agreement with a municipality. In this for example fifty-year-long agreement, the manufacturer could undertake the installation, commission, maintenance and replacement of the wind turbine (when it will reach at the end of its lifetime) and the profits could be shared between the manufacturer and the independent power producer and/or the municipality. How such a business model will be viable in other sectors? Is implementing CE in other sectors financially sustainable?

CONCLUSION

The goal of this work was to investigate behind the known theory and propose an innovative concept/business approach focusing on circularity and reusability. It was revealed that end-users since they are familiar with this approach in the automotive industry, they would prefer to buy a service instead of undertaking the cost of ownership. A number of weights are removed of the customers' shoulders, such as the responsibility of maintenance and the value loss. Therefore, it is the manufacturing companies that is required to change mentality, since it seems that the end-users are ready to deal with the upcoming changes.

The research was concluded by a calculation of the lowered amount of new products needed to be produced and the lessened CO_2 emissions (on in-bound logistics) if Kvik decided to introduce CE as part of their business strategy. It was found that Kvik could reduce their production with 19% and their CO_2 emissions through in-bound logistics could be reduced by 40%.

NOTES

1. Figures 8.1 through 8.6 available online at https://rowman.com/ISBN/9781793625021/Sustainable-Engineering-for-Li.

REFERENCES

Aizcorbe, A., and Starr-McCluer, M. (1997). Vehicle ownership, purchases, and leasing: Consumer survey data. *Monthly Labour Review*, 120, pp. 34–40.

Amui, L., Jabbour, C., de Sousa Jabbour, A., and Kannan, D. (2017). Sustainability as a dynamic organizational capability: A systematic review and a future agenda toward a sustainable transition. *Journal of Cleaner Production*, 142, pp. 308–322.

Ausrød, V., Sinha, V., and Widding, Ø. (2017). Business model design at the base of the pyramid. *Journal of Cleaner Production*, 162, pp. 982–996.

Avgoustaki, D. D., and Xydis, G. (2020). Indoor vertical farming in the urban nexus context: Business growth and resource savings. *Sustainability*, 12(5), 1965. doi: 10.3390/su12051965

Aviles, A., Bottcher, J., and Xydis, G. (2020, June 13). Solar-powered golf buggies charging on the road. *Energy Sources, Part A: Recovery, Utilization, and Environmental Effects*. doi: 10.1080/15567036.2020.1792589

Bennett, E., and Graedel, T. (2000). "Conditioned air": Evaluating an environmentally preferable service. *Environmental Science and Technology*, 34(4): 541–545.

Business Dictionary. (2017). *How has this term impacted your life?* [online]. Available at: http://www.businessdictionary.com/definition/manufacturing.html [Accessed 15 Nov. 2017].

CEFIC. (2011). *Guidelines for measuring and managing CO_2 emission from freight transport operations*. [online]. Available at: http://www.cefic.org/Industry-support /Responsible-Care-tools-SMEs/5-Environment/Guidelines-for-managing-CO2-em issions-from-transport-operations/ [Accessed 21 Apr. 2018].

Danmarks Statistik. (2016). *Finansieringsselskaber 2015*. [online]. Available at: http://www.dst.dk/da/Statistik/nyt/NytHtml?cid=22551 [Accessed 13 Dec. 2017].

Duguay, C., Landry, S., and Pasin, F. (1997). From mass production to flexible/ agile production. *International Journal of Operations & Production Management*, 17(12), pp. 1183–1195.

Duray, R. (2002). Mass customization origins: Mass or custom manufacturing? *International Journal of Operations & Production Management*, 22(3), pp. 314–328.

European Union. (2015). *Sustainable Development in the European Union*. Luxembourg: Publications Office of the European Union. Available at: https://ec .europa.eu/eurostat/documents/3217494/6975281/KS-GT-15-001-EN-N.pdf

European Union Council. (1999). *Council Directive 1999/31/EC on the Landfill of Waste*. Brussels. Available at: https://eur-lex.europa.eu/legal-content/en/TXT/?ur i=CELEX%3A31999L0031

Frausing, E. (2018). Kvik production, *Freight*, Interview (Personal Contact).

García-Olivares, A., and Solé, J. (2015). End of growth and the structural instability of capitalism—From capitalism to a symbiotic economy. *Futures*, 68, pp. 31–43.

Haanstra, W., Toxopeus, M., and van Gerrevink, M. (2017). Product life cycle planning for sustainable manufacturing: Translating theory into business opportunities. *Procedia CIRP*, 61, pp. 46–51.

Hart, C. (1995). Mass customization: Conceptual underpinnings, opportunities and limits. *International Journal of Service Industry Management*, 6(2), pp. 36–45.

Hawken, P., Lovins, A., and Lovins, L. (2000). *Natural Capitalism*. New York: Little, Brown and Company.

Hu, S. (2013). Evolving paradigms of manufacturing: From mass production to mass customization and personalization. *Procedia CIRP*, 7, pp. 3–8.

Intlekofer, K., Bras, B., and Ferguson, M. (2010). Energy implications of product leasing. *Environmental Science & Technology*, 44(12), pp. 4409–4415.

Korhonen, J., Honkasalo, A., and Seppälä, J. (2017). Circular economy: The concept and its limitations. *Ecological Economics*, 143, pp. 37–46.

Koroneos, C. J., Polyzakis, A., Xydis, G. A., Stylos, N., and Nanaki, E. A. (2017). Exergy analysis for a proposed binary geothermal power plant in Nisyros Island, Greece. *Geothermics*, 70, pp. 38–46. doi: 10.1016/j.geothermics.2017.06.004

Koscis, G., and Xydis, G. (2019). Repair process analysis for Wind Turbines equipped with Hydraulic Pitch mechanism on the U.S. market in focus of cost optimization. *Applied Sciences*, 9, p. 3230. doi: 10.3390/app9163230

Lausselet, C., Cherubini, F., Oreggioni, G., del Alamo Serrano, G., Becidan, M., Hu, X., Rørstad, P., and Strømman, A. (2017). Norwegian waste-to-energy: Climate change, circular economy and carbon capture and storage. *Resources, Conservation and Recycling*, 126, pp. 50–61.

Li, K., and Xu, S. (2015). The comparison between trade-in and leasing of a product with technology innovations. *Omega*, 54, pp. 134–146.

Mannering, F., Winston, C., and Starkey, W. (2002). An exploratory analysis of automobile leasing by US households. *Journal of Urban Economics*, 52(1), pp. 154–176.

Mathews, J. (2011). Naturalizing capitalism: The next great transformation. *Futures*, 43(8), pp. 868–879.

Nanaki, E. A., Kiartzis, S., and Xydis, G. A. (2020, Jan). Are only demand-based policy incentives enough to deploy electromobility? *Policy Studies*. doi: 10.1080/01442872.2020.1718072

Nanaki, E. A., Koroneos, C. J., Xydis, G. A., and Rovas, D. (2014). Comparative environmental assessment of Athens urban buses—Diesel, CNG and biofuel powered. *Transport Policy*, 35, pp. 311–318. doi: 10.1016/j.tranpol.2014.04.001

Nanaki, E. A., Xydis, G. A., and Koroneos, C. J. (2015). Electric vehicle deployment in urban areas, indoor and built environment (special issue: urban sustainability). *Indoor and Built Environment*, 25(7), pp. 1065–1074. doi: 10.1177/1420326X15623078

Nikas, E., Sotiropoulos, A., and Xydis, G. A. (2018). Spatial planning of biogas processing facilities in Greece: The sunflower's capabilities and the waste-to-bioproducts approach. *Chemical Engineering Research and Design* (Special Issue: Energy Systems Engineering), 131, pp. 234–244. doi: 10.1016/j.cherd.2018.01.004

Pallant, J. L., Sands, S., and Karpen, I. O. (2020). The 4Cs of mass customization in service industries: A customer lens. *Journal of Services Marketing*, Article in Press.

Pedersen, L. (2018). Kvik production, *Freight*, Interview (Personal Contact).

Polanyi, K. (2014). *The Great Transformation*. Boston, MA: Beacon Press, p. 136.

Rasmussen, N. B., Enevoldsen, P., and Xydis, G. (2020). Transformative multi-value business models: A bottom-up perspective on the hydrogen-based green transition for modern wind power cooperatives. *International Journal of Energy Research*. doi: 10.1002/ER.5215

Ritzén, S., and Sandström, G. (2017). Barriers to the circular economy—Integration of perspectives and domains. *Procedia CIRP*, 64, pp. 7–12.

Saunders, M., Lewis, P., and Thornhill, A. (2015). *Research Methods for Business Students*. 1st ed. New York: Pearson Education.

Singh, J., and Ordoñez, I. (2016). Resource recovery from post-consumer waste: Important lessons for the upcoming circular economy. *Journal of Cleaner Production*, 134, pp. 342–353.

Solomon, S. (2008). *Climate Change 2007*. Cambridge, MA: Cambridge University Press, p. 996.

Sousa-Zomer, T., Magalhães, L., Zancul, E., and Cauchick-Miguel, P. (2017). Exploring the challenges for circular business implementation in manufacturing companies: An empirical investigation of a pay-per-use service provider. *Resources, Conservation and Recycling*, 135, pp. 3–13.

Trocchia, P., and Beatty, S. (2003). An empirical examination of automobile lease vs finance motivational processes. *Journal of Consumer Marketing*, 20(1), pp. 28–43.

Tukker, A. (2015). Product services for a resource-efficient and circular economy—A review. *Journal of Cleaner Production*, 97, pp. 76–91.

Tukker, A., and Tischner, U. (2006). Product-services as a research field: Past, present and future. Reflections from a decade of research. *Journal of Cleaner Production*, 14(17), pp. 1552–1556.

Xydis, G., Nanaki, E., and Koroneos, C. (2013). Exergy analysis of biogas production from a municipal solid waste landfill. *Sustainable Energy Technologies and Assessments*, 4, pp. 20–28. doi: 10.1016/j.seta.2013.08.003

Zadnik Stirn, L., Gornik Bučar, D., and Hrovatin, J. (2016). Examination of decision factors in the process of buying kitchen furniture using conjoint analysis. *Drvna Industrija*, 67(2), pp. 141–147.

Chapter 9

Should Canada Pursue or Support Offshore Wind Development?

Sarah Nichol, Lindsay Miller, and Rupp Carriveau

INTRODUCTION

Wind energy is a clean and cost-effective solution to meet increasing energy demands while facing climate change challenges. In areas where terrestrial space is becoming scarce, offshore installations are becoming popular due to significant energy potential associated with higher wind speeds and vast offshore areas. Since 2013, the global offshore wind market has grown on average 24% each year, resulting in total installations of 29.1 GW at the end of 2019 (Lee and Zhao 2020). Most installations are off the coast of European countries with the United Kingdom and Germany accounting for 59% of total global installations. However, in the past few years many installations have been added in Asia, with China now accounting for 24% of total global offshore wind installations. It is predicted that over 205 GW of new offshore wind will be installed over the next decade (Lee and Zhao 2020). While Canada is a massive nation rich in energetic natural resources, there are currently no offshore wind turbines in Canadian waters.

Although offshore growth is not as prominent in North America as it is in Europe, there have been some recent developments of offshore wind in the United States. The first U.S. offshore farm came online in December 2016 and there are now over 25,000 MW of offshore wind in the U.S. pipeline (Lee and Zhao 2020). The first offshore project in the Great Lakes is currently under development and is expected to be operational by the end of 2022 (Lake Erie Energy Development Corporation 2020). There has been significant policy support for offshore wind in the United States over the past few years and growth of offshore wind in the United States is expected to continue as renewable energy targets increase while costs of offshore projects decline.

Research has revealed that Canada has significant resource potential for offshore wind and has illustrated the areas where installations would deliver the greatest benefits (Doubrawa et al. 2015; Barrington-Leigh and Ouliaris 2017). However, such opportunities must be economic to be pragmatic. Offshore wind projects face higher costs of installation and operating and maintenance due to access challenges and the need for large crane vessels along with the potential for environment impacts on marine ecology. In Canada, a country with vast terrestrial area available for onshore installations, the business case for offshore installations is very challenging. Compared to its onshore counterpart, offshore wind has greater resource potential due to stronger and steadier winds, and has the opportunity for installation of large turbines, up to 8 MW, while minimizing noise and visual impacts. Currently, several offshore wind projects are in the planning stages in Newfoundland, Nova Scotia, New Brunswick, Prince Edward Island, and in British Columbia. Although Ontario has been identified as having significant potential for offshore installations, a moratorium is in place on all offshore wind projects and has been since 2011, halting offshore development plans for the time being.

Canada is also a nation rich in experience, expertise, and capabilities in the energy sector. Offshore global capacity is expected to approach 234 GW by 2030 (Lee and Zhao 2020) and Canadian businesses can capitalize on the CAD$350 billion that will be invested to support this growth through provision of services and expertise to the sector. Some Canadian companies have realized this and have recently become involved in the offshore scene. Exciting business opportunities for Canadian companies will be available to those who can strategically position themselves and closely follow the global offshore progress.

There are many opportunities for Canada in the offshore energy markets; however, the best option is not clear. The purpose of this study is to determine how Canadian stakeholders can best position themselves to become key players in the rapidly growing offshore wind industry. With the potential for such large investments in this industry, and the major benefits of offshore wind technology, it is important to consider if and how Canada should be involved. The study pulls information from many sources to lay out the barriers, drivers, support, and courses of action available to Canada regarding the offshore wind industry. The comparisons of these to other offshore wind projects and markets will provide insight to how Canada should proceed in the industry. With various opportunities available for Canada's involvement, a plan must be determined, in which the areas of best practices are considered. Immediate action is required if Canada is to become involved in the offshore scene, as developments are rapidly expanding, and the United States is beginning to search for contributors to their offshore wind supply chain.

This paper first identifies drivers and barriers to offshore wind development with a focus on relevance to Canadian projects. Then an overview of North American policies relating to offshore wind is summarized, considering both the United States and Canada. Following this, global offshore activities, with a focus on recent U.S. developments are explored, followed by Canadian developments. Furthermore, this paper presents an examination of options for prioritizing the next steps for Canada to capitalize on offshore wind opportunities.

OFFSHORE WIND DRIVERS AND BARRIERS

Offshore Wind Drivers

Offshore wind has several advantages over its onshore counterpart, the greatest being that offshore wind has a greater resource potential. Winds blow stronger and more consistently over water than over land, and are greatest during the day when energy demands are highest (Government of Canada). When winds are varying offshore, they are more easily predictable, with the potential to be forecasted days in advance (Marine Renewables Canada 2018). Higher wind speeds imply higher productivity which can partially offset the higher capital costs. Higher capacity factors are also a benefit of offshore wind compared to onshore (Kaldellis and Kapsali 2013). If installed sufficiently far from shore, offshore installations can avoid many of the conflicts posed by land-based wind projects, including visual impacts, noise, and shadow flicker. In areas with scarce land, offshore wind can provide an obvious option for additional energy production with minimal terrestrial footprint. Furthermore, there is virtually no limit on the size of the turbines since there are no concerns with road limit restrictions as there are with onshore transportations (Breton and Moe 2008). Due to the ease in transportation of large structures by water, offshore turbines can be much larger than onshore, allowing for much greater energy production potential (Marine Renewables Canada 2018).

Offshore wind energy is a clean and sustainable source of energy with the potential for very large-scale projects to significantly reduce the use of carbon-emitting energy supplies. Under the Paris Agreement, Canada committed to reducing its greenhouse gas emissions to 30% below the levels of 2005 by the year 2030. The country also is generating a plan for a net-zero emissions future by the year 2050 (Environment and Climate Change Canada 2020). The integration of offshore wind into Canada's energy sector could help the country meet these goals as targeted.

Likewise, smaller-scale offshore wind farms could power remote communities, reducing the use of diesel fuel (Marine Renewables Canada 2018).

There are over 292 remote communities in Canada which are often supplied with diesel-dependent imported energy. Many of these communities have access to renewable energy resources, including waters and offshore wind. Indigenous communities pose an ideal community space for these wind farms to be located. Indigenous participation in the clean energy sector within Canada has rapidly increased over the past two decades. Over the next few years, continuous participation in clean energy projects by Indigenous communities is expected as they move away from the use of diesel-reliant energy (Henderson and Sanders). Smaller-scale offshore wind project installations within these communities can provide a transition to cleaner energy increased community self-reliance, as well as job and skill developments (Marine Renewables Canada 2018). Remote communities, especially coastal areas, also suffer from frequently compromised transmission lines and unreliable energy (Marine Renewables Canada 2018). The integration of offshore wind in these areas would also be beneficial, with the point of use generation improving the reliability of energy and reducing maintenance on transmission lines (Marine Renewables Canada 2018).

Offshore Wind Barriers

Although there are many advantages to offshore wind turbines, there are also disadvantages that pose as challenges to implementation. A large barrier in implementing offshore wind turbines are the associated construction, maintenance, and transmission costs, as they are significantly greater than the costs associated with onshore wind turbines (Musial and Ram 2010). The weather, winds, waves, and water currents experienced by offshore turbines require a different approach in terms of technology, support structures, electrical infrastructure resulting in higher investments in towers, foundations, and underwater cabling. Construction of offshore wind farms also requires use of materials that can resist the corrosive marine environment. Costs of offshore wind in North America are about twice that of onshore wind (Breton and Moe 2008; Delucchi and Jacobson 2010). A major challenge is access to offshore site for maintenance and repairs. Offshore turbine repairs can be five to ten times more expensive than onshore, largely due to the need for expensive crane vessels (Zaaijer and Bussel 2001). Furthermore, waiting for acceptable weather conditions to perform repairs can inflict turbine downtime and lost production will negatively impact the economics of the project. Another challenge facing offshore development is the possibility of shortage of materials, such as the number of vessels available for construction and repair as more projects begin development. Lastly, offshore turbines are exposed to harsher environmental conditions such as wave and current loading, and possible icing and corrosion. Project developers have also cited difficulties in securing construction contracts under which risks are carried by the contractor due to unwillingness to bear the significant geotechnical and weather risks associated with offshore development (Hartley 2006). In this case, the developer would have to assume this risk, which could also make project financing more difficult. Dramatic cost reductions have been achieved in the last couple of years. The global average of LCOE for offshore wind according to BNEF has dropped over 67% since 2012, now sitting at \$84/MWh. This is expected to continue to drop to \$58/MWh by 2025 (Lee and Zhao 2020). The reasons for the significant cost declines include growing investor confidence and the introduction of new GW sized projects.

In addition to monetary disadvantages, there are also environmental concerns that have been raised surrounding offshore development and studies have been commissioned (Isaacman and Daborn 2011). The predominant concerns include possible negative impacts from noise, lubricant spills, and electromagnetic fields from cables on marine life (Hammar et al. 2014). Additional potential impacts include disturbance of fish breeding, disturbance of communication between species, and reductions in habitat size due to operational noise emissions (Kaldellis et al. 2016). Despite these being raised as possible concerns, studies have demonstrated that species were only affected during the pile-driving operations of construction and that low-intensity wind turbine noise during operation posed little possibility of causing hearing impairment of fish (Wahlberg and Westerberg 2005). Other works have suggested that there may be beneficial ecological impacts through provision of habitat to marine species as offshore turbines can act as artificial reefs (Kaldellis et al. 2016). Environmental concerns can largely be addressed through proper siting and technology and material selection. Design and material innovations aimed at simplifying manufacturing and installation processes are underway and are expected to result in fewer impacts on marine life. Although impacts can be very site specific, in general, offshore developments pose low impacts to the environment especially when compared with other energy generation (Kaldellis et al. 2016).

Fleet challenges can also have substantial impacts on project timing and costs. In the United States, the "Jones Act" restricts the "transportation of merchandise by water" between "points in the United States" to qualified U.S. vessels. This act complicates the construction, operation, and maintenance of offshore wind farms in the United States as there will be a general requirement for merchandise to be moved between a U.S. port and towers attached to the seabed, or between towers (Imhof 2018). The purpose-built wind turbine installation vessels used in European developments (and not built in the United States) cannot be used in U.S. waters. The offshore U.S. market could eventually support development of multiple purpose-built vessels, with preliminary plans for such underway. In the meantime, developers are relying on resourceful solutions to suit their situation (Imhof 2018). Even without regulatory restrictions, projects face challenges in accessing a fleet of highly specialized vessels that are required for offshore wind installations.

The most notable barrier to offshore wind development in Canada is the moratorium that is currently in place in Ontario, a province with significant offshore wind potential. In February 2011, Ontario announced a moratorium on all offshore wind development in the province, citing visual impacts amongst the major concerns (Spears 2013). Several large projects were affected by this decision including Superior Array (650 MW), Trillium Power Wind I (480 MW), Trillium Power Wind II (740 MW), Great Lakes Array (1600 MW), Erie Wind Energy (4000 MW), and Wolfe Island Shoals (300 MW). Both Windstream Energy and Trillium Power Wind had wind turbine projects planned for Lake Ontario when the government brought down the moratorium. Again in 2017, the province has signalled that the moratorium will remain in place for some time, despite being hit with a $25 million penalty under the North American Free Trade Agreement for stalling the Windstream project (Ferguson 2016). The province cites the need for more research on noise and environmental issues, despite the five government-commissioned studies that have taken place since 2011. The province also plans to monitor the impacts of the Icebreaker project in the Great Lakes.

Adequate terrestrial space is also a barrier to offshore development. In some countries, space is becoming scarce for onshore wind turbines (Government of Canada), and therefore, offshore developments provide a suitable option. With terrestrial space still available in Canada, making a business case for offshore development can be challenging. Furthermore, there are technical challenges surrounding the installation. Many of the potential offshore sites in North America are located in deeper waters than those installed globally, requiring new technologies for installation. Stronger wind and larger wave and ice loadings would also have to be considered. Especially in the Great Lakes, a major concern is the freshwater icing that occurs. Areas with the greatest water depth are ice-free, but the wind and wave conditions may be harsher, and different floating offshore wind technologies would need to be explored. Deep-water prototypes have successfully been installed in Norway (220 m depth) and Italy (110 m depth) demonstrating the ability of the technologies to be adapted for North American waters (Kaldellis et al. 2016).

Furthermore, vessel size is more restricted in the Great Lakes than other areas, as vessels larger than 23.7 meter cannot enter the St. Lawrence Seaway locks (Smith et al. 2016). If specialized vessels are created for the construction and decommissioning of offshore turbines in Canada, they may not be as capable as those present in Europe (Smith et al. 2016). The build-out of high voltage transmission has also been identified as a potential challenge to offshore wind development in the Great Lakes.

OFFSHORE WIND POLICIES IN NORTH AMERICA

Different policies exist within North America providing various levels of support for the development of offshore wind.

Offshore Wind Policies in the United States

Recent policy support for offshore wind in the United States will further support this sector. In May 2017 Maryland awarded the Public Service Commission's first offshore wind renewable energy credits (ORECs) to two projects (Delony 2017). The ORECs were awarded to Deepwater Wind's 120 MW Skipjack Project, expected to be operational in 2022, and U.S. Wind's 270 MW project, expected to be operational in 2023 (US Wind). Each of these projects will receive an OREC for US$131.93/MWh for twenty years which will significantly improve the economics and directly allow them to be built (Delony 2017).

Massachusetts committed their electricity distribution companies to produce 1600 MW of offshore wind energy by June 2027 (Massachusetts Clean Energy Center). Also, in 2017, Massachusetts released the nation's first competitive solicitation for commercial-scale offshore wind (Stori 2018) which resulted in proposals from three developers. Different funding programs were also created to support research and development and workforce training on offshore wind. Many states have continued their support of offshore wind development. Rhode Island, Massachusetts, and New York have all passed policies to support private investments in offshore wind projects in 2017 (Polefka 2018), and since then have continued to support the industry.

The National Offshore Wind Research and Development Consortium was established in 2018 to collaborate with industry to advance research and development focused on improving technical, supply chain, social, or economic barriers faced by the offshore wind industry in the United States (National Offshore Wind Research and Development Consortium). Resulting from a solicitation for R&D projects in 2019 the Consortium awarded $17.3 million in funding. Recently (August 2020), another $9 million of funding was announced to be awarded to projects focused on the current challenge areas of offshore wind in the United States. These challenge areas are focused improving the costs of development, barriers to deployment and growth of the offshore wind industry (National Offshore Wind Research and Development Consortium). An non-American company can be awarded this funding, permitting that the proposal is focused on benefiting the U.S. offshore wind industry (National Offshore Wind Research and Development Consortium).

Offshore Policy in Canada

Policy developments in Canada are necessary to shape the future of offshore energy. In February 2018, the Canadian federal government proposed how it plans to regulate offshore renewable energy projects. Part 5 of the Canadian Energy Regulator Act defines how the government will regulate offshore renewable energy projects and includes a commission to regulate activities, an authorization regime, and a process by which proponents can apply for an authorization (House of Commons of Canada 2018). The proposed act also specifies impact assessment timelines and liability and financial requirements for these projects.

Canada's minister of natural resources also announced the Emerging Renewable Power Program, which will provide funding intended to reduce risks and associated costs with emerging renewable energy technologies not yet established in Canada. This could include offshore wind (Froese 2018). The funds are also intended to help emerging sectors navigate regulatory issues and ultimately provide Canada with a more diverse set of clean energy technologies.

Another encouraging development for offshore wind in Canada came from the announcement that Marine Renewables Canada will officially be including offshore wind energy in their mandate (Obermann 2018). The organization, previously focused on wave and tidal energy, now recognizes that there are significant overlaps between these technologies and offshore wind when it comes to supply chain, regulatory issues, and operating environment and plans to focus on the roles that each can play in Canada's low-carbon future.

The 2018 State of the Art Sector Report for Marine Renewable Energy in Canada laid out some funding support available within Canada for potential marine renewable energy projects, which could include offshore wind. The Green Infrastructure Fund is focused on clean energy development in remote communities. The Clean Growth in Natural Resources Program can support clean technology research and development in Canada's natural resource sectors in energy (Marine Renewables Canada 2018).

Comparing policies and incentives generated by the United States, Canada is not receiving the same support as the United States in the offshore wind industry. This poses a major barrier to offshore developments in Canada.

OFFSHORE WIND DEVELOPMENTS

Offshore Wind Outside of North America

At the end of 2019, there was 29.1 GW of installed offshore wind around the globe, accounting for 5% of total global wind capacity (see figure 9.1).[1] Although Europe remains the leader in total offshore wind installations, hosting

75% of total global capacity (see figure 9.2), activity in Asia, specifically China, has been continuously increasing throughout the past few years (see figure 9.1). China has now installed 24% of the world's offshore wind capacity, the third highest of any country after the United Kingdom (with 33% of total global capacity) and Germany (with 26% of total global offshore wind capacity) (see figure 9.3). 2019 was the greatest year in history for the offshore wind industry, as over 6.1 GW of new capacity was installed (Lee and Zhao 2020). China installed the largest capacity of new offshore wind in 2019 at 2.4GW, the United Kingdom installed 1.8GW and Germany installed 1.1 GW, followed by Denmark and Belgium (see figure 9.4) (Lee and Zhao 2020).

Development of Offshore Wind in the United States

The first U.S. offshore wind farm came online in December 2016. The 30 MW, five-turbine Block Island project installed off the coast of Rhode Island has secured a twenty-year PPA to sell its electricity. The project will connect Block Island to the mainland grid for the first time resulting in a 40% reduction in island electricity rates (Orsted). The Block Island Farm is linked to the New England electricity grid by National Grid's new sea2shore submarine transmission cable system. The project, built on-time and on-schedule and employing hundreds of local workers, was a success story for offshore wind. The successful implementation of offshore turbines in U.S. waters has paved the way for more U.S. offshore developments.

The United States has many other projects in the pipeline, the majority of which are proposed to be located in Massachusetts, New Jersey, and North Carolina (Norton and Pitts 2017). The Skipjack wind farm is an example of a planned project in the United States (Orsted). The 120 MW wind farm is located in the Atlantic Ocean, 19.5 miles from its closest point in Maryland. The project plans to come online by 2023 (Orsted).

Plans for the largest combined offshore wind and energy storage project in the world, Revolution Wind off of the coast of Massachusetts, are also underway. The project was approved for a twenty-year PPA for 400 MW of clean energy (Orsted 2019). The project is predicted to be coming online in 2023 (Orsted 2019). Support for offshore wind in the United States is evident from the U.S. DOE's roadmap for wind power which envisages offshore wind providing 2% of U.S. electricity demand by 2030 and 7% by 2050 (U.S. Department of Energy (DOE) 2015).

Offshore Wind in the Great Lakes: Icebreaker Project

The first Great Lakes offshore wind project is currently being developed by the Lake Erie Energy Development Corporation (Lake Erie Energy

Development Corporation 2020). The demonstration scale Icebreaker project, sponsored by the U.S. Department of Energy, will be the first freshwater offshore wind project in North America and will produce electricity to power over 7,000 homes. The 20.7 MW project will be located 13 kilometer off the coast of Cleveland, Ohio in Lake Erie in water 18 m in depth. The distance from shore was selected based on available meteorological tower wind data from the Cleveland Water Intake Crib. Icebreaker Wind will interconnect with the Cleveland Public Power transmission system with secured rights to participate in the PJM market.

Findings from the Environmental Assessment have demonstrated negligible or minor, short-term impacts across resource areas. Preliminary analysis and monitoring shows minor, short-term adverse impacts on fish and avian species as well as lake water quality. Negligible impacts were reported for drinking water supply and quality, insects, and aquatic and terrestrial protected species. The project is also anticipated to create 500 jobs and to result in community, environmental, and health benefits. The project plans to be operational by the end of 2022 (Lake Erie Energy Development Corporation 2020). This project is of particular interest to developers considering projects in Canada since it is within Lake Erie, bordering the United States and Canada. If successful, this project can settle environmental, economic, and regulatory concerns and catalyse future offshore development in the Great Lakes.

The Great Lakes have approximately 136 GW of technical resource capacity (Musial et al. 2016). This number assumes that none of the water above 60 meters in depth would be feasible for development due to icing concerns. An NREL study identified the Great Lakes as being relatively low cost due to the high-quality wind resource and general absence of extreme meteorological ocean events such as strong winds and waves (Beiter et al. 2015). Tides and water currents are negligible in the Great Lakes, and wave heights are much smaller than those in open ocean conditions (Smith et al. 2016). Levelized cost of energy values for offshore development in the Great Lakes are estimated to be as low as \$75/MWh by 2027 (Beiter et al. 2015). The least-cost locations found are in Lake Erie, Lake Michigan, and Saginaw Bay in Lake Huron. These locations have strong wind resources that are close to shore. A wind atlas was recently developed for the Great Lakes to provide a blueprint for offshore wind resource assessment effort (Doubrawa et al. 2015). The wind atlas demonstrated that for most of the province, turbines at a height of 90 m would be met with speeds greater than 8 m/s and would achieve a mean energy density of greater than 314 W/m^2 (Doubrawa et al. 2015). In addition, Great Lakes shorelines are also peppered with coal plants, offering a low-cost opportunity for grid connection. Developers and regulators will be closely following the impact of the

Icebreaker project to gauge the feasibility of future developments in the Great Lakes.

Development of Offshore Wind in Canada

The offshore wind energy potential in the Atlantic and Pacific Oceans and in the Great Lakes is enormous. Boasting the longest coastlines in the world, offshore wind potential in Canada has been estimated at over 500 TWh/year (Barrington-Leigh and Ouliaris 2017) with British Columbia and Ontario possessing by far the most potential. The Great Lakes hold enough energy potential to power the entire country and the winds of Lake Erie alone could meet over 10% of our electricity needs by 2030 (Lake Erie Energy Development Corporation 2020). Barrington-Leigh and Ouliaris (2017), set out to determine the optimal locations for offshore wind in Canada. The authors defined high potential offshore winds sites as those with large areas of water (> 25 km^2) with average depths less than 30 meter and average wind speeds (at a height of 80 meter above sea level) greater than 8 m/s. Based on those criteria, high potential sites were identified around four main areas: off the coast of British Columbia, the Great Lakes, the Gulf of St. Lawrence, and off the coast of Nova Scotia near the Bay of Fundy (Barrington-Leigh and Ouliaris 2017). Although the maximum potential of these sites would ultimately be affected by shipping lanes, environmental concerns, and winter freezing over of freshwater (Great Lakes) (Barrington-Leigh and Ouliaris 2017), there is no doubt that significant amounts of clean energy could be supplied by offshore turbines in Canadian waters.

Planned Projects in Canada

To date, there are no operational offshore wind turbines in Canada. There are, however, several offshore wind projects in the planning stages. In British Columbia, the Naikun Wind Energy Group recently sold their offshore wind project to Northland Power Inc. (Northland Power 2020). The 396 MW wind project will be located in the Hecate Strait between Gwaii and Prince Rupert, off the coast of British Columbia. This project has faced many setbacks and is still in the early stages of development (Northland Power 2020).

On the East Coast of Canada, Beothuk Energy Inc. has several projects in the pipeline. Beothuk's C$466 million, 180 MW St. Georges Bay offshore wind project is proposed to be located 30 kilometer offshore of Bay St. George, with water depths averaging 40 meter (Dawes 2016). The proposed site has some advantages such as being outside of shipping lanes, away from bird migration routes, and near Emera's Maritime Link Transmission line. The company has also proposed a C$4 billion offshore wind farm 20 kilometer from Yarmouth, Nova Scotia (Dawes 2016). The intention for this project is to sell the energy generated from 120 turbines to New England through a 370 kilometer subsea cable called the Can-Am Link (2015). Progress with this system has been delayed due to developments in other plans belonging to the project partners (The Telegram 2017). Beothuk has additional projects planned for St. Ann's Bay, Nova Scotia, Burgeo Banks, Newfoundland, and Prince Edward Island.

There was also interest in Ontario for implementing wind farms throughout the Great Lakes; however, those plans were terminated in 2011 when the Ontario Government put a stop to all projects until more scientific studies are completed (Blackwell 2011). Originally, Ontario became an attractive location for offshore wind farm development due to the combination of the Great Lakes wind conditions, and the "Feed-in-Tariff" program that incentivized renewable power (Blackwell 2011). The government of Ontario stated the reasons for cancelling projects for offshore wind farms in the Great Lakes included limited scientific research in freshwater wind turbines, even though there are hundreds of offshore developments in ocean waters throughout the world (Blackwell 2011), and a few freshwater farms (Blackwell 2011).

Canadian Companies on the Global Offshore Scene

Canada has vast experience in offshore and marine industries which makes the nation a strong candidate for offshore wind energy development projects. There are many synergies between offshore oil and gas and the offshore sectors suggesting transferability of knowledge and products between the two, as has already been demonstrated in the United Kingdom (Roberts et al. 2014). Similarities between the offshore oil and gas and wind industries are present at each step of the development process. The project management experience of offshore oil companies is important in the offshore wind industry due to their understanding of the harsh working environments, maintenance needs and coordination of working in offshore waters. Offshore oil and wind projects require similar expertise for the design and supply of equipment including substation structures, foundations, secondary steel, array cables and cable protection. Developers are especially likely to hire expertise from the oil and gas industry for cable installations, as it is considered a high-risk and challenging operation (Duncan 2017). Many of the foundations considered in the offshore wind industry have been previously deployed in the offshore oil and gas industry, although under different loadings (Duncan 2017). This understanding of the different foundation types can be transferred to determine the necessary design for a certain project's water depth, seabed, and loading conditions. Companies with experience in the construction and installation of offshore oil and gas platforms are also very well suited to apply their knowledge to offshore wind projects. Much of the equipment and skills required in the handling and installation of components is similar. Both industries have many standards, certifications, and maintenance requirements in common as well (Duncan 2017). Due to the mentioned synergies above and many more, knowledge and skills obtained from Canada's years of experience in the offshore oil and gas industry can be transferred to applications in the offshore wind industry.

In order to introduce offshore wind to Canadians in the current offshore field, educational and training programs can be redirected for studies relating to offshore wind, and current graduates and employees in the offshore sector of Canada can be trained specifically on development of offshore wind turbines. Canadian companies are recognising these and capitalizing on global offshore opportunities.

Northland Power, a Canadian company based out of Toronto, Ontario, is 85% owner and 100% operator of the 332 MW Nordsee One offshore wind farm in the German North Sea (Northland Power). The wind farm is expected to produce an annual output of more than 1,300 gigawatt-hours of electrical energy, enough to supply the equivalent of approximately 400,000 German households. Northland also acquired 60% ownership of the 600 MW Gemini offshore wind project, located 85 kilometers off the Netherlands coast. This project was completed ahead of schedule, in April 2017, and under budget. It is expected to generate enough clean and renewable energy to meet the needs of 1.5 million people in the Netherlands, and reduce the country's CO_2 emissions by 1.25 million tons per year (Northland Power). Northland also has 100% ownership of the 252 MW DeBu offshore project in Germany, 77 kilometer from the Nordsee One project.

Enbridge, a Canadian energy company is also actively involved in the global offshore scene. In November 2015, Enbridge entered the offshore scene with its acquisition of a 24.9% stake in the 400 MW Rampion project developed by E.ON (Enbridge 2018). The project consists of 116 Vestas 3.45 MW turbines located in the English Channel 13–20 kilometer south of Brighton (Enbridge 2018). In February 2017, Enbridge also acquired 50% ownership (with the remaining 50% retained by EnBW) in the 497 MW Hohe Sea Offshore Wind Project located in the North Sea, 98 km off of the German coast (EnBW). The project will receive a twenty-year fixed pricing contract under the German government's offshore wind incentive structure. Recognizing the vast potential for offshore wind, Enbridge has also partnered with EDF Energy on the development, construction, and operation of three French offshore wind farms which would collectively produce 1,428 MW of power (Yu 2016). The Saint-Nazaire Offshore Wind Project, 50 percent owned by Enbridge, is expected to enter service in 2022 (Enbridge).

With global offshore estimated to approach 234 GW by 2030 (Lee and Zhao 2020) comes exciting business opportunities for companies who can provide expertise and services to this sector. There are many ways in which Canadian companies can capitalize on the approximately CAD$350 billion that will be invested in capital and operational expenditures over this time-frame. Project development, engineering services, software and modelling, research and education, marine vessels, geotechnical and geophysical ser-vices, turbine supply, electrical systems, and port facilities are some more of the areas of the offshore supply chain where Canadian companies could play a role (Stapleton 2017).

Current Best Practices for Offshore Wind in Canada

Three scenarios for Canada's involvement in the offshore wind industry are considered: (1) install in Canadian waters and sell power to Canada or export power, (2) do not install in Canada and instead focus on exporting Canadian expertise and supporting the international supply chain, or (3) it is not yet beneficial for Canada to be involved in the offshore scene in any form.

Canada has a very high potential for the development of offshore wind along its coastlines, however, significant installations are not very likely in the foreseeable future due to a low-value proposition of such development. Across the country, terrestrial space is largely available, which makes it difficult to accept the development of more expensive offshore wind rather than continuing to expand onshore wind energy. The Canadian federal and provincial governments have not provided many policies, funding, or sup-port for offshore wind, which is a large barrier in its development across the country. In the United States, government support, especially at the state levels, has been the key in the development of necessary research, testing and project plans. In addition, Canada already has a relatively clean energy grid, with many options for other clean energy sources that can be integrated further. Also, there has not been enough research completed on the effects of offshore wind in freshwater environments, which rules out any opportunities for development in the Canadian waters of the Great Lakes for the foresee-able future.

In the West Coast of Canada, such as in British Columbia, there is a high population density along the shoreline which may demonstrate offshore wind to be useful in the future when developments become more feasible. Offshore wind could also provide many benefits for Canadian communities, especially remote and coastal communities currently dependent on diesel-resources for energy, or with compromised energy infrastructure. The west coast and remote communities could be potential suiters of offshore wind use in Canada. However, challenges are expected in these areas due to the deep waters requiring floating technology and the potential icing which has not been largely tested in offshore installations. Floating wind turbines have been previously installed in global projects; however, the technology is newer, more expensive and has higher risk during extreme winds and waves. Without funding and support from government or private sources, or further understanding of deep water and icing challenges on floating offshore wind technologies, installing offshore wind along the Canadian West Coast or in remote communities is not recommended for sale of energy to Canada. This is due to the uncertainties and lack of research in these water types. However, if supported as a research project, installations in these areas are important to further the understanding of challenges being faced by this technology.

Along the Eastern coast of Canada, there are very long shorelines; however, the population density along them is not very high. Considering the low population density along the Eastern shoreline of Canada, building up extremely large offshore wind farms may not be economical or necessary to meet energy needs. Due to this, installations of offshore turbines in Eastern Canada to only provide energy to Canadian populations is not recommended at this time. However, if the Canadian government provided more support, funding and research findings, developments in Canadian waters could become viable. Additionally, following the results of U.S. offshore wind projects could provide valuable insight to optimizing projects in similar waters within Canada. The United States has many projects in their pipeline for installations in waters along the Eastern Coastline, and an overwhelming interest as well as high targets for integration of offshore wind energy.

By following the successes of the upcoming U.S. projects, especially those in similar waters, Canadian developers can use the lessons learned from the United States to confidently implement infrastructure along the Canadian Eastern Coastline. Although the Canadian population density in this area may not be large enough for the full potential of offshore wind production, these developments could sell their power to bordering U.S. states, where population densities are high (Government of Canada 2020). This would allow Atlantic Canada to become a provider of offshore wind energy to Eastern U.S. states looking to improve their green economy (Natural Resources Canada 2017). With the United States actively building up its offshore wind capacity along their bordering shorelines with Canada, economics will play a large role in the ability for Canada to sell power generated by future offshore wind farms on its coast to the United States. Even with the head start in offshore wind development and support within the United States, this may be possible, as the United States has developed restrictions on the use of foreign wind towers (Leone 2012; Ministry of Foreign Affairs of Denmark the Trade Council 2020) and vessels within their developments. These restrictions in infrastructure use will increase the costs of energy projects within the United

States (Efstathiou 2018). With support from the Canadian Federal government an energy contract could be generated between the US and Canada,
allowing the sale of Canadian generated offshore wind energy at a competitive price to bordering U.S. locations. This could help in creating an economical plan for offshore wind project installations in Canada. However, further
developments in both the United States and Canada are necessary before
development in Canadian waters to export energy sales becomes a practical
next step for Canadian offshore wind developers.

In the current state, installation of offshore wind in Canadian waters for
energy sales to Canada is not a feasible option for Canadian developers.
Following the development of government support and further understanding of floating wind in waters similar to Canadian waters, installing offshore
wind in Canadian waters for the sale of energy to bordering U.S. states is recommended. In the present, Canada should contribute and benefit from global
offshore wind developments by exporting expertise to projects elsewhere.
This is the recommended best course of action for Canadian involvement in
the offshore wind industry currently. Canadian expertise of the offshore oil
and gas industry can be valuable when transferred over to offshore wind, as
there are many synergies between the two industries. This has been proven
by the many successful partnerships of international offshore wind projects
with Canadian developers to date.

Future work should focus on the development of a supply chain database
for offshore wind in Canada (similar to that provided by Marine Renewables
Canada for other marine technologies) to identify expertise and capabilities
within the country and match these to global project needs.

CONCLUSIONS

The future for offshore wind energy in Canada is very promising. Global
and North American capacity will continue to increase over the next decade,
spurred by climate targets, decreasing costs, and resource quality. This will
require massive investments in the offshore sector but will also generate
many great benefits. Canada has abundant resources, experience, and inter-
ested developers to support offshore wind development. There are several
advantages that offshore wind boasts over its onshore counterpart, including
stronger and more predictable winds, larger turbine capacity, and aesthetic
appeal of installations far from shore and their suitability for powering remote
communities. These advantages rationalize why offshore wind energy is an
important consideration for Canadian waters and companies alike. However,
the barriers to installation of offshore turbines make the decision for involvement in offshore installations difficult. Offshore turbines are much more

costly than their onshore counterpart, and although they save on terrestrial space, Canada has expansive land available currently. These findings are in parallel with Bull (2014) who also found that due to the abundance of land and alternative lower-cost energy sources, it is unlikely for offshore wind to take off in Canada. However, this statement did not consider the potential for sales of energy generated by offshore wind to the United States. In addition, in European waters, where offshore wind is the most popular, turbines are developed for much more shallow waters than there are in Canada, and do not face as high weather concerns including icing, and therefore uncertainties in the necessary technologies are present. It was also found that the lack of government support and related policies is a major obstacle for the growth of offshore installations in Canada. Studying past offshore wind developments and areas advanced in this industry, it is clear that support is necessary for initial integration to be feasible.

Although Canada is not currently poised to launch significant offshore developments, with the global growth of the industry projected to drastically increase, and Canadian businesses should be seeking out new opportunities in this sector. Providing expertise on international offshore wind projects is the recommended first step for Canada's involvement in the offshore scene. Immediate contribution to international projects can ensure leadership in the offshore wind supply chain internationally, and especially in North America, which is important due to the rapid growth of the industry. Recently, this has been stressed by government of Canada (2020) as U.S. offshore wind projects advance, seeking international expertise. Canadian developers should also consider involvement in manufacturing offshore wind turbine parts, a potential that Clean Energy Canada (2015) considers Canada to be currently missing out on.

Following Canada's entry to the offshore wind supply chain, it is recommended that the next step would be to explore the potential of implementing offshore wind in Canadian waters to export power to the United States. This second recommendation could be implemented after closely following the outcomes of United States offshore wind developments. Eventually, with more support, funding, and declining technology costs, offshore wind energy may become a feasible option to meet the energy needs of Canada as well.

Canada is in a great position to benefit from lessons learned from the United States, global experiences, and from the improvement and maturation of technology and growing investor confidence. The Icebreaker Wind Project in Lake Erie will provide a valuable case to evaluate North American specific impacts and gauge feasibility of future projects. In relation to Canadian policies regarding offshore wind, while the moratorium remains in place in Ontario, development will be limited. At the federal level, stable, coordinated policy is needed to offset high initial costs and drive development.

Finally, it is recommended that future work should develop a supply chain database to identify expertise in the offshore industry in Canada and match it to the needs of global offshore wind projects.

ACKNOWLEDGMENTS

The authors would like to acknowledge the support of the Natural Sciences and Engineering Research Council of Canada and the valuable inputs from the Offshore Renewables Canada Working Group participants.

NOTES

1. Figures 9.1 through 9.4 available online at https://rowman.com/ISBN/9781793625021/Sustainable-Engineering-for-Life-Tomorrow.

REFERENCES

Barrington-Leigh C, Ouliaris M (2017) The renewable energy landscape in Canada: A spatial analysis. *Renewable and Sustainable Energy Reviews* 75: 809–819. https://doi.org/10.1016/j.rser.2016.11.061

Beiter P, Musial W, Kilcher L, et al. (2017) An assessment of the economic potential of offshore wind in the United States from 2015 to 2030. Technical Report NREL/TP-6A20-67675, March 2017.

Blackwell R (2011) Ontario stops offshore wind power development. *The Globe and Mail*. theglobeandmail.com. Accessed 1 Oct 2020.

Breton S-P, Moe G (2008) Status, plans and technologies for offshore wind turbines in Europe and North America. https://doi.org/10.1016/j.renene.2008.05.040

Bull J (2014) The United States should prioritize offshore wind—But should Canada? *UBC Sauder*. https://www.sauder.ubc.ca/news/us-should-prioritize-offshore-wind-should-canada. Accessed 8 Nov 2020.

Clean Energy Canada (2015) *Tracking the Energy Revolution*. Vancouver, BC. trackingtherevolution.ca/canada

Dawes T (2016) Beothuk Energy awards contracts to build Newfoundland offshore wind farm. https://www.cantechletter.com/2016/06/beothuk-energy-awards-contracts-build-newfoundland-offshore-wind-farm/. Accessed 1 Oct 2020.

Delony J (2017) Maryland awards renewable energy credits to two offshore wind projects. *Renewable Energy World*. https://www.renewableenergyworld.com/2017/05/12/maryland-awards-renewable-energy-credits-to-two-offshore-wind-projects/. Accessed 30 Sep 2020.

Delucchi MA, Jacobson MZ (2010) Providing all global energy with wind, water, and solar power, part II: Reliability, system and transmission costs, and policies. *Energy Policy* 39: 1170–1190. https://doi.org/10.1016/j.enpol.2010.11.045

Doubrawa P, Barthelmie RJ, Pryor SC, et al. (2015) Satellite winds as a tool for offshore wind resource assessment: The Great Lakes Wind Atlas. *Remote Sensing of Environment* 168: 349–359. https://doi.org/10.1016/j.rse.2015.07.008

Duncan A (2017) *Oil and Gas Diversification: Offshore Wind.* Cricklade, Swindon, UK. https://www.norwep.com/content/download/30687/221159/version/1/file/BVG+Alan+Duncan+Offshore+Wind+Workshop.pdf

Efstathiou J (2018) Trump's import tariffs will make U.S. wind power more expensive—Renewable energy world. *Bloomberg.* https://www.renewableenergyworld.com/2018/10/03/trumps-import-tariffs-will-make-us-wind-power-more-expensive/#gref. Accessed 1 Oct 2020.

Enbridge (2018) *A Colossal Achievement: Rampion Offshore Wind Farm Now Fully Operational.* Enbridge Inc. https://www.enbridge.com/stories/2018/november/rampion-offshore-wind-farm-fully-operational. Accessed 30 Sept 2020.

Enbridge Saint Nazaire Offshore Wind Project. https://www.enbridge.com/projects-and-infrastructure/projects/saint-nazaire-offshore-wind-project. Accessed 30 Sept 2020.

EnBW. EnBW Hohe See and Albatros wind farms: One joint project in the North Sea. https://www.enbw.com/renewable-energy/wind-energy/our-offshore-wind-farms/hohe-see/

Environment and Climate Change Canada (2020) Canadian environmental sustainability indicators: Progress towards Canada's greenhouse gas emissions reduction target. www.canada.ca/en/environment-climate-change/services/environmental-indicators/progresstowards-canada-greenhouse-gas-emissions-reduction-target.html. Accessed 29 Sept 2020.

Ferguson R (2016) Ontario extends moratorium on offshore wind turbines. https://www.thestar.com/news/queenspark/2016/12/22/ontario-extends-moratorium-on-offshore-wind-turbines.html. Accessed 30 Sept 2020.

Froese M (2018) Canada supports new renewable sources, including offshore wind. https://www.windpowerengineering.com/canada-supports-new-renewable-sources-including-offshore-wind/. Accessed 30 Sept 2020.

Government of Canada. Canada's adoption of renewable power sources: Energy market analysis—Emerging technologies. https://www.cer-rec.gc.ca/nrg/sttstc/lctrct/rprt/2017cnddptnrnwblpwr/mrgngtchnlgs-eng.html. Accessed 29 Sept 2020.

Government of Canada (2020) U.S. offshore wind power holds promise for Canada's ocean technology firms. https://www.tradecommissioner.gc.ca/canadexport/0005061.aspx?lang=eng. Accessed 9 Nov 2020.

Hammar L, Wikström A, Molander S (2014) Assessing ecological risks of offshore wind power on Kattegat cod. *Renewable Energy* 66: 414–424. https://doi.org/10.1016/j.renene.2013.12.024

Hartley K (2006) The winds of change—EPC contracts for offshore wind farms. https://www.mondaq.com/uk/industry-updates-analysis/39948/the-winds-of-change--epc-contracts-for-offshore-wind-farms? Accessed 30 Sept 2020.

Henderson C, Sanders C. *Powering Reconciliation: A Survey of Indigenous Participation in Canada's Growing Clean Energy Economy.* Ottawa, ON: LUMOS Clean Energy, Oct 10, 2017.

House of Commons of Canada (2018) Bill C-69 An Act to enact the Impact Assessment Act and the Canadian Energy Regulator Act, to amend the Navigation

Protection Act and to make consequential amendments to other Acts. *Parliament of Canada.* https://www.parl.ca/DocumentViewer/en/42-1/bill/C-69/first-reading. Accessed 30 Sept 2020.

Imhof JF (2018) What does the Jones act mean for offshore wind? https://www.mar inelink.com/news/offshore-jones-does434220. Accessed 30 Sept 2020.

Isaacman L, Daborn G (2011) *Pathways of Effects for Offshore Renewable Energy in Canada.* Wolfville, NS: FERN.

Kaldellis JK, Apostolou D, Kapsali M, Kondili E (2016) Environmental and social footprint of offshore wind energy: Comparison with onshore counterpart. *Renewable Energy* 92: 543–556.

Kaldellis JK, Kapsali M (2013) Shifting towards offshore wind energy—Recent activity and future development. *Energy Policy* 53: 136–148. https://doi.org/10.1 016/j.enpol.2012.10.032

Lake Erie Energy Development Corporation (2020) The project: Icebreaker wind. http://www.leedco.org/index.php/about-icebreaker. Accessed 30 Sept 2020.

Lee J, Zhao F (2020) *Global Offshore Wind Report 2020.* Brussels, Belgium.

Leone S (2012) DOC imposes tariffs on Chinese wind towers. http://www.alte nergystocks.com/archives/2012/05/doc_imposes_tariffs_on_chinese_wind_towers /. Accessed 1 Oct 2020.

Marine Renewables Canada (2018) *Marine Renewable Energy in Canada 2018 State of the Sector Report.*

Massachusetts Clean Energy Center Offshore Wind (MassCEC). https://www.mass-cec.com/offshore-wind. Accessed 30 Sept 2020.

Ministry of Foreign Affairs of Denmark The Trade Council (2020) US ITC to impose duties on imported wind turbine towers from Canada, South Korea and Vietnam. *DK Wind Energy Advisory.* https://www.offshorewindadvisory.com/industry-news /20-07-us-itc-antidumping-canada-skorea-vietnam/. Accessed 1 Oct 2020.

Musial W, Heimiller D, Beiter P, et al. (2016) *2016 Offshore Wind Energy Resource Assessment for the United States,* NREL/TP-5000-66599.

Musial W, Ram B (2010) *Large-Scale Offshore Wind Power in the United States: Assessment of Opportunities and Barriers.* NREL/TP-500-40745.

National Offshore Wind Research and Development Consortium. NOWRDC announces availability of $9 million to advance offshore wind innovation. https ://nationaloffshorewind.org/news/press-release-innovation-in-offshore-wind-solic itation-1/. Accessed 30 Sept 2020.

Natural Resources Canada (2017) Western newfoundland becoming a "rock solid" investment for offshore wind energy. *Government of Canada.* https://www.nrcan .gc.ca/cleangrowth/20417. Accessed 29 Sept 2020.

Northland Power (2020) Northland power provides business update on COVID-19 situation and reiterates its 2020 financial guidance. https://www.northlan dpower.com/Investor-Centre/News--Events/Press_Releases.aspx?NewsID=6556. Accessed 30 Sept 2020.

Northland Power Nordsee One (Offshore Wind). https://www.northlandpower.com /What-We-Do/Operating-Assets/Wind/Nordsee-One.aspx. Accessed 1 Oct 2020.

Norton G, Pitts B (2017) 4 Emerging trends in U.S. offshore wind technologies. https ://www.energy.gov/eere/articles/4-emerging-trends-us-offshore-wind-technolo gies. Accessed 30 Sept 2020.

Obermann E (2018) Marine renewables Canada to focus on synergies with offshore wind. *Marine Renewables Canada.* https://marinerenewables.ca/marine-renew ables-canada-to-focus-on-synergies-with-offshore-wind/. Accessed 30 Sept 2020.

OffshoreWindBiz (2015) Beothuk proposes 1GW wind farm off Nova Scotia. *OffshoreWindBiz.* https://www.offshorewind.biz/2015/12/22/beothuk-proposes -1gw-wind-farm-off-nova-scotia/. Accessed 1 Oct 2020.

Orsted (2019) Rhode island regulators approve revolution wind power contract. https://us.orsted.com/news-archive/2019/05/rhode-island-regulators-approve-revol ution-wind-power-contract. Accessed 30 Sept 2020.

Orsted about Skipjack wind farm. https://skipjackwindfarm.com/about-skipjack-wind -farm. Accessed 30 Sept 2020.

Orsted our offshore wind projects in the U.S. https://us.orsted.com/wind-projects. Accessed 30 Sept 2020.

Polefka S (2018) State policies can unleash U.S. commercial offshore wind development. *Center for American Progress.* https://www.americanprogress.org/is sues/green/reports/2017/09/18/439078/state-policies-can-unleash-u-s-commercial -offshore-wind-development/. Accessed 1 Oct 2020.

Roberts A, Blanch M, Weston J, Valpy B (2014) UK offshore wind supply chain: Capabilities and opportunities, BVG Associates.

Smith G, Drunsic M, Reynolds P, Whitmore A (2016) *Assessment of Offshore Wind Farm Decommissioning Requirements.* Ontario Ministry of the Environment and Climate Change, Document No: 800785-CAMO-R-06, Montreal, QC.

Spears J (2013) Ontario's off-shore wind turbine moratorium unresolved two years later. https://www.thestar.com/business/economy/2013/02/15/ontarios_offshore_ wind_turbine_moratorium_unresolved_two_years_later.html. Accessed 30 Sept 2020.

Stapleton E (2017) Offshore wind energy development: Supply chain identification and capacity within Newfoundland and Labrador.

Stori V (2018) Will 2018 be another good year for U.S. offshore wind? *Renewable Energy World.* https://www.renewableenergyworld.com/2018/01/09/will-2018-be -another-good-year-for-us-offshore-wind/. Accessed 30 Sept 2020.

The Telegram (2017) Beothuk energy project delayed: Offshore Nova Scotia mega-wind farm pushed down the priority list for parent company Copenhagen Offshore Partners. https://www.thetelegram.com/business/beothuk-energy-project-delay ed-offshore-nova-scotia-mega-wind-farm-pushed-down-priority-list-for-parent -company-copenhagen-offshore-partners-133703/. Accessed 1 Oct 2020.

U.S. Department of Energy (DOE) (2015) *Wind Vision: A New Era for Wind Power in the United States.* Washington, D.C: U.S. Department of Energy (DOE).

U.S. Wind Maryland Offshore Wind Project. http://www.uswindinc.com/maryland-o ffshore-wind-project/. Accessed 30 Sept 2020.

Wahlberg M, Westerberg H (2005) Hearing in fish and their reactions to sounds from offshore wind farms. *Marine Ecology Progress Series* 288: 295–309. https://doi.org/10.3354/meps288295

Yu V (2016) Enbridge's interest in European offshore wind gathers velocity with close of acquisition. https://www.enbridge.com/stories/emf-french-offshore-wind-project. Accessed 30 Sept 2020.

Zaaijer MB, Van BG (2001) Reliability, availability and maintenance aspects of large-scale offshore wind farms, a concepts study. *Proceeding of MAREC 2001*, Newcastle, UK.

Chapter 10

Analyzing the Renewable Energy Sources of Nordic and Baltic Countries with MCDM Approach

Fazıl Gökgöz and Engin Yalçın

INTRODUCTION

Climate change is a global phenomenon that is affiliated with economics, environment, and energy. Global warming will continue unless more precautions are taken to prevent it. This phenomenon could cause extreme rises in temperature, an increase in the sea level, storms, and extreme precipitation (Sharifi, 2020). These situations would also have a negative effect on economies across the world. Thus, decreasing greenhouse gas (GHG) emission is of utmost importance in order to mitigate climate change (Zheng et al., 2019). Climate change mitigation is a crucial strategy in preventing climate change in the long term, as it focuses on the causes of climate change. Focus on the energy supply is at the core of climate change mitigation endeavors, as it is the largest source of climate change. In this context, GHG emissions from fossil fuels pose a threat to the environment, so renewable energy sources (RESs) are indispensable sources in mitigating climate change. RESs are sources that may be replenished in a short time period and can be derived from the sun and other natural processes. RESs include sources such as bioenergy, thermal, geothermal, and hydroelectric, all with nearly zero emissions. RESs are generally used for the production of heat, electricity, and vehicle fuels (Elum & Momodu, 2017).

Today, fossil fuels comprise roughly 80 percent of the global energy demand, and they are the main sources of GHG emission all over the world (Wennersten et al., 2015). RESs are highly efficient tools for dealing with environmental problems. RESs produce extremely low levels of the emissions that contribute to climate change compared to fossil fuels (Wang et al., 2018). In this respect, energy efficiency improvements and reduction of the use of carbon-intensive energy sources are beneficial precautions for mitigating climate change on the production side. These efforts, however, should not be limited to the production side. The consumption of energy sources should also be reduced to reach climate change mitigation targets (Christis et al., 2019). On the other hand, the level of climate change is heavily determined by the scale of economic activities (Leimbach et al., 2017). Thus, we analyze climate change mitigation and economic performance in an integrated manner via multi-criteria decision-making (MCDM) methods.

MCDM methods are viable operational research approaches in analyzing the performance of alternatives with many criteria. MCDM methods can be deterministic, stochastic, or fuzzy (Pohekar & Ramachandran, 2004). So far, many different MCDM methods have been introduced to cope with decision-making problems *via* various criteria. There are circumstances when the best possible alternatives should be ranked and evaluated depending on the established criteria. MCDM methods are suitable approaches for dealing with these problems (Yazdani et al., 2019). Energy-planning problems have drawn the attention of researchers for a long time, and MCDM methods have been used in energy-planning problems recently as they are more suited to the domain (Løken, 2007).

In this study, we aim to analyze economic performance and climate change mitigation in Nordic and Baltic countries over the period of 2013–2017. The Nordic area includes Denmark, Norway, Sweden, Finland, and Iceland. The Baltic states contain the countries of Estonia, Latvia, and Lithuania. We select the Nordic and Baltic region for our analysis for a few reasons. Initially, we should mention that Nordic and Baltic countries invest heavily on RES to mitigate climate change (Irandoust, 2016; Siksnelyte et al., 2019). Promoting RES-oriented long-term policies has also paved the way for economic development and sustainability in Nordic and Baltic countries (Ardakani et al., 2018; Miskini et al., 2020). Their strategic breakthroughs and policies enable them to be benchmarks for other countries, possibly setting an example for other countries that aim to increase economic performance and climate change mitigation.

The remainder of this study is organized as follows. The following section presents the literature review on energy performance studies. Then the research methods are introduced. The last section of the study presents comparative empirical analysis, conclusions, and policy implications.

LITERATURE REVIEW

Many papers were analyzed to evaluate energy performance on international and national levels. These studies were generally based on data envelopment analysis (DEA) and MCDM methods. DEA is a nonparametric approach to assessing alternatives (Gökgöz, 2010). The following paragraphs detail some of these studies analyzing the energy performance of the countries and provinces.

Halkos and Petrou (2019) assessed the environmental efficiency via a DEA approach over the period of 2008–2014. They focused on municipal solid waste (MSW) production, employment rate, capital formation, gross domestic product (GDP), Sulphur oxide (SOx), nitrogen oxide (NOx), and GHG emissions for European Union (EU) countries, concluding that Sweden was efficient for all of the investigated periods. Cucchiella et al. (2018) evaluated the 2020–2030 European strategy of EU countries employing a zero sum gains DEA approach. They evaluated country performance in the context of economic, social, and environmental indicators. GDP was the economic indicator of the analysis, and population was the social aspect of the analysis. GHG emissions and renewable energy consumption were the environmental indicators. They concluded that most EU countries demonstrate an increasing trend of efficiency. The average efficiency level for EU countries has increased over the last ten years.

Baležentis et al. (2016) evaluated the climate change mitigation and sustainable energy performance of the Lithuanian economy. They used gross value added as the desirable output, while the inputs were the total hours worked by employees, capital stock, and emission-related energy use, and CO_2 and N_2O emissions were the undesirable outputs. They concluded that energy consumption and carbon emissions decreased over the investigated period. This situation mainly stems from decreasing energy intensity.

Wang et al. (2020) evaluated the economic performance and climate change mitigation of thirty-nine global cities using DEA. They used population as input, while their desirable output was GDP, and three different GHG emission variables were used as undesirable outputs. They formed four models based on natural, managerial disposability, basic, and super efficiency approaches. They concluded that the energy mix and economic structure should be reinforced together in low-performing countries.

Škrinjarić (2020) evaluated the circular economy of some European countries over the period of 2010 to 2016 via Grey Relational Analysis (GRA). Škrinjarić (2020) used energy recovery per capita, recycling material per capita, gross investment percentage of GDP, employment rate in the circular economy, circular material use, and the number of patents in relation to recycling and secondary raw materials. Škrinjarić (2020) concluded that Germany, Denmark, the Netherlands, and France were found to be the best-performing countries. On the other hand, Slovakia, Romania, Portugal, and Greece were found to be the worst-performing countries.

Mi et al. (2017) evaluated the climate change mitigation performance of Chinese regions via a TOPSIS approach. In their paper, GDP, population, primary energy supply, CO_2 emission and energy intensity targets are selected as performance indicators derived from the Climate Change Performance Index (CCPI). They concluded that the southern part of China performs better than other regions.

RESEARCH METHODS

Shannon Entropy Method

It is essential to compute the relative criterion importance to evaluate the alternatives in MCDM problems. There are two classifications of weighting approaches: subjective and objective weights. Objective weighting methods depend on mathematical calculation, without taking into consideration the decision maker's preferences (Wang & Lee, 2009). In our study, we employ entropy method to avoid the human judgment variable of subjective weighting approaches. Shannon Entropy is a method that enables decision-makers to weigh criteria in an objective manner. Shannon Entropy was later adapted into various research fields. MCDM is one of those fields in which entropy approach is employed by researches in the decision science (Fedajev et al., 2020). This entropy method is widely employed in MCDM methods due to its objective significance coefficient (Maghsoodi et al., 2018). The entropy approach is one of the most reliable weighting methods for MCDM methods (Arya & Kumar, 2020).

The entropy notion was first introduced in thermodynamics, before Shannon (1948) adapted this notion into information theory. In information theory, the entropy concept is regarded as the rate of the irregularity in processes. Thus, this indicator may be employed to evaluate the inhomogeneity of the indices. If the standard values of different indicator indices differ considerably, the entropy value will be lower (Yang et al., 2018).

Assume that there are m options to appraise n criteria. $(x_{ij})_{m*n}$ is the first decision matrix. The initial step of the entropy approach is to establish a decision matrix (Mavi et al., 2016).

$$X = \begin{bmatrix} a_{11} & \cdots & a_{12} & \cdots & a_{1n} \\ \vdots & \ddots & \vdots & \ddots & \vdots \\ a_{21} & \cdots & a_{22} & \cdots & a_{2n} \\ \vdots & \ddots & \vdots & \ddots & \vdots \\ a_{m1} & \cdots & a_{m2} & \cdots & a_{mn} \end{bmatrix} \quad (10.1)$$

where every row signifies the mth option while every column denotes the n-th criteria. X denotes the decision matrix.

The normalization procedure is displayed as follows:

$$p_{ij} = x_{ij} / \sum_{i=1}^{m} x_{ij} \qquad (10.2)$$

where, p_{ij} denotes the normalized decision values.

The information entropy for each index is calculated as follows:

$$E_j = -(\ln m)^{-1} \sum_{i=1}^{m} p_{ij} \ln p_{ij} \qquad (10.3)$$

where, m is the number of options, E_j is the entropy value of an evaluated criterion and ln denotes the natural logarithm.

The weight acquired via information entropy is shown as follows:

$$w_j = \left(1 - E_j\right) / (n - \sum_{j=1}^{n} E_j) \qquad (10.4)$$

where, n denotes the number of criteria, w_j is the objective weight based on the entropy concept, where;

$$0 \leq w_j \leq 1 \text{ and } \sum_{j=1}^{n} w_j = 1. \qquad (10.5)$$

The VlseKriterijumska Optimizacija I Kompromisno Resenje (VIKOR) Method

The VIKOR approach is an MCDM method introduced by Opricovic (1998) to provide compromises to solve problems. This approach focuses on measuring the ideal distances. It is also a suitable technique for ranking alternatives with conflicting criteria (Xu et al., 2017, p. 260). The VIKOR method, which was proposed as an alternative to the ELECTRE approach, is based on the nearness to the ideal solution. The VIKOR approach introduces an aggregating function reflecting nearness to the ideal alternative (San Cristóbal, 2011). The principal objective of the VIKOR approach is to obtain an optimized solution among the alternatives with conflicting criteria (Jahan et al., 2011).

The steps of the VIKOR approach are presented below (Ploskas & Papathanasiou, 2019; Xu et al., 2017).

Step 1: Computation of the best and the worst criteria values:

The initial step of the VIKOR approach is to calculate the best and worst criteria values.

$$f_j^+ = max_i f_{ij}, \quad f_j^- = min_i f_{ij}, j = 1,2,\ldots\ldots,n \text{ for benefit criteria} \quad (10.6)$$

$$f_j^- = min_i f_{ij}, \quad f_j^- = max_i f_{ij}, j = 1,2,\ldots\ldots,n \text{ for cost criteria} \quad (10.7)$$

where f_j^+, f_j^- denote the best and worst criteria values, respectively.

Step 2. Computation of utility and regret measures:

In the VIKOR approach, utility and regret measures are calculated as follows:

$$S_i = \sum_{j=1}^{n} w_j \left(f_j^+ - f_{ij} \right) / \left(f_j^+ - f_j^- \right) \quad i = 1,2,\ldots\ldots.m \quad (10.8)$$

$$R_i = max_j \left\{ w_j \left(f_j^+ - f_{ij} \right) / \left(f_j^+ - f_j^- \right) \right\} \quad i = 1,2,\ldots\ldots.m \quad (10.9)$$

where w_j is the weight of criteria obtained through entropy method. S_i and R_i are the utility and regret measures respectively.

Step 3. Computation of VIKOR index values:

The VIKOR index for every alternative is calculated as follows:

$$Q_i = v \frac{\left(S_i - S^+ \right)}{\left(S^- - S^+ \right)} + \frac{\left(1 - v \right)\left(R_i - R^+ \right)}{\left(R^- - R^+ \right)} \quad i = 1,2,\ldots\ldots.m \quad (10.10)$$

where $S^+ = min_i S_i$, $S^- = max_i S_i$, $R^+ = min_i R_i$, $R^- = max_i R_i$, vi is the coefficient of the group utility function and it is generally regarded as 0.5, which is a compromised way of evaluating the alternatives. S_i and R_i denote the utility and regret measures respectively. Q_i values are calculated to rank the alternatives in the VIKOR method.

Step 4. Ranking the alternatives and acquiring evaluation results:

The options are ranked in increasing order. The alternative with the minimum Q_i value is found to be the best option because it complies with the following two requirements.

Step 5. Obtaining compromise solution

If the option $(A)^{(1)}$ satisfies the following requirements, $(A)^{(1)}$ is regarded as the optimum option. This step reflects the stability of the analysis.

R1. Acceptable advantage

$$Q(A^{(2)}) - Q(A^{(1)}) \geq 1/(n-1) \qquad (10.11)$$

where $(A)^{(1)}$ is the finest ranked option, $(A)^{(2)}$ is the second finest option, and n denotes the number of options.

R2. Acceptable stability in decision making: The option $(A)^{(1)}$ should be ranked as the finest alternative by S or/and R computations.

DATA, VARIABLES, AND PROFILES OF NORDIC AND BALTIC COUNTRIES

In our study, we aim to analyze economic performance and climate change mitigation in Nordic and Baltic countries over the period of 2013–2017. We aim to analyze Nordic and Baltic countries since they provide an important portion of global wealth, and they demonstrate high performance in climate change mitigation. Our data covers the period of 2013–2017, which are the most recent data available. Our dataset is retrieved from the Eurostat (EU Statistical Bureau) database, which provides technical indicators for European countries. We analyze the economic performance and climate change mitigation of Nordic and Baltic countries via the VIKOR approach, which is highly relevant among MCDM methods.

Economic and Climate Change Mitigation Profiles of Nordic and Baltic Countries

Energy plays a key role in production processes and the world economy. Intensive energy consumption, particularly stemming from fossil fuels, is a serious threat to the environment, since it causes more GHG emission compared to RESs. An increased amount of GHG emission has adverse effects on human health and the environment (Wood & Roelich, 2019; Khanali et al., 2020). Roughly 70% of worldwide GHG emissions are due to the combustion of fossil fuels. The main alternative for decreasing GHG emissions is to use RESs, which cause nearly no GHG emissions. RESs are regarded as sustainable and clean options for the environment (Manish et al., 2006). Thus, countries are strongly recommended to focus on RES-oriented policies to mitigate climate change.

Economic growth is affected by energy consumption and climate change (Sarkodie et al., 2020). In this respect, economic performance and climate change mitigation are interrelated and should be handled in an integrated manner. To this end, we aim to analyze the economic performance and

Table 10.1   Unit of Data and Descriptions of Technical Indicators

Technical Indicators	Unit of Data	Definitions of Technical Indicators
Final energy consumption	Million tons of oil equivalent (TOE)	This indicator includes all energy consumed by final consumers. This total consumption is comprised of industry, transport, households, services and agriculture.
GHG emissions	Thousand tons CO_2 equivalent	This indicator shows the total amount of CO_2, nitrous oxide, methane, and other emissions caused by burning of carbon-containing fuels. It is one of the principal reasons of global warming.
Energy intensity	Kilograms of oil equivalent (KGOE) per thousand euro	This indicator shows how much energy is necessary to generate a unit of GDP.
The share of energy derived from RES in final energy consumption	Percentage	This indicator shows the share of RES use in final energy consumption.
GDP	Constant Prices USD per head	This indicator is the monetary amount of all finished products and services made within a country during a particular time

climate change mitigation of Nordic and Baltic countries via the VIKOR method. In this part, we introduce the technical indicators used to analyze countries in the study: final energy consumption, GHG emissions, energy intensity, the share of energy derived from RES, and GDP. The unit of data and the definitions of technical criteria are presented in table 10.1.

The descriptive statistics of technical indicators for the environmental and climate change mitigation performances of EU members are presented in table 10.2.

In our study, we use GDP, which is an economic output indicator frequently employed to reflect the growth of an economy (Kim, 2013). The average GDP amount in Nordic and Baltic countries in the period of 2013 to 2017 is presented in figure 10.1.[1]

The average GDP amount in Nordic and Baltic countries increased by 8% during this period. Energy intensity contributes to economic growth in the Nordic and Baltic countries. A 13% decrease in energy intensity signifies improvement in energy efficiency, and this progress reflects the economic growth of countries (Irandoust, 2016). Energy intensity is used to describe consumption compared to output. A higher energy intensity level reflects that more energy is needed to generate a unit of output. Energy intensity may also be interpreted as an indicator to rate the benefit and cost situation for the environment and energy (Tan & Lin, 2018).

Table 10.2 Descriptive Statistics* of Technical Indicators

Years	Indicators	Minimum	Maximum	Average	Standard Deviation
2013	Final energy consumption	2.9	31.9	13	10.5
	GHG emissions	10492.6	57389.6	24609.7	16337.4
	Energy intensity	72	415.3	200.7	110.1
	The share of energy derived from RES	22.6	71.6	42.2	17.6
	GDP	23181.3	59317.9	40126.8	11893.9
2014	Final energy consumption	2,8	31,1	12,7	10,2
	GHG emissions	11511,5	52828,2	24240,5	14017,3
	Energy intensity	67,8	406,7	191,8	106,4
	The share of energy derived from RES	23,5	70,4	43,5	17,3
	GDP	23845,2	59815,1	40714,7	11715,2
	Final energy consumption	2.8	31.7	12.9	10.3
2015	GHG emissions	11746.4	53788.4	24068.1	14337.5
	Energy intensity	66.3	374.4	181.4	96.4
	The share of energy derived from RES	25.7	70.2	44.2	16.6
	GDP	24833.6	60368.9	41609.3	11768.9
	Final energy consumption	2.8	32.1	13.2	10.5
	GHG emissions	8497	56830	24575.9	16019
2016	Energy intensity	66	338.7	179.5	90.9
	The share of energy derived from RES	25.6	70.1	44.4	16.7
	GDP	25505.9	60479.7	42597.3	11786.9
	Final energy consumption	2.9	32.3	13.3	10.5
	GHG emissions	9647.1	52839.2	23729.3	14331.5
2017	Energy intensity	65.11	345.9	176.7	88.9
	The share of energy derived from RES	26	71.6	45.7	16.6
	GDP	26719.8	61404	43737.3	11493.2

Source: Eurostat.
* *The descriptive statistics are calculated by authors depending on the Eurostat data.*

Energy efficiency and decarbonization are two important targets involved in the EU 2020 Energy strategy. Energy intensity ratio, which measures the efficiency of economies, is a frequently used concept in evaluating economic performance (Velasco-Fernández et al., 2018). The average energy intensity level in Nordic and Baltic countries in the period of 2013 to 2017 is presented in figure 10.2.

The average energy intensity level in Nordic and Baltic countries decreased by 13% over during the investigated period. This result indicates efficiency improvement in their economies, which is a positive sign for climate change mitigation, as well (Irandoust, 2016; Miskinis et al., 2020). The average energy consumption in Nordic and Baltic countries in the period of 2013 to 2017 is presented in figure 10.3.

Cold weather conditions increase energy consumption in Nordic and Baltic countries (Santamouris, 2016). Despite this shortcoming, their energy consumption remains lower in comparison to most other EU countries. This result heavily depends on the energy efficiency of the Nordic and Baltic countries. Fossil fuels damage climate change mitigation endeavours, causing a high amount of GHG emission into the atmosphere (Johnsson et al., 2019). Renewable energy use is regarded as an efficient tool to mitigate climate change (Chen et al., 2020). Nordic countries are pioneers among EU countries in terms of renewable energy (Aslani et al., 2013; Siksnelyte et al., 2019). The share of RES in final energy consumption in Nordic and Baltic countries in the period of 2013 to 2017 is presented in figure 10.4.

One of the most significant causes of global warming are GHG emissions, which mainly derive from fossil fuels. To this end, decreases in GHG emissions are quite important to mitigating climate change (Cardenas et al., 2016). The average GHG emissions in Nordic and Baltic countries in the period of 2013 to 2017 are presented in figure 10.5.

GHG Emissions in Nordic and Baltic countries have remained stable over the investigated period. The majority of GHG emissions derive from energy sources. Thus, the share of renewable energy in final energy consumption is regarded as a highly important policy for mitigating climate change (Babatunde et al., 2017).

EMPIRICAL ANALYSIS

In this study, economic performance and climate change mitigation is assessed in Nordic and Baltic countries via the VIKOR method in the period from 2013 to 2017. The weight values attributed to criteria via the entropy method in the period are displayed in table 10.3.

We analyze the economic performance and climate change mitigation of Nordic and Baltic countries via the VIKOR method. The VIKOR method results in the period from 2013 to 2017 are presented in table 10.4.

Table 10.3   **Weight Values Attributed to Criteria via Entropy Method**

Years	Final Energy Consumption	GHG Emissions	Energy Intensity	The share of energy derived from RES	GDP
2013	0.412	0.247	0.178	0.105	0.056
2014	0.437	0.205	0.196	0.103	0.057
2015	0.445	0.218	0.185	0.093	0.056
2016	0.429	0.258	0.169	0.090	0.052
2017	0.446217	0.237	0.175	0.090	0.050

Source: Authors' calculations.

According to VIKOR method results, Sweden has been found to be the best-performing country among the Nordic and Baltic countries. Lithuania has been found to be worst-performing country over the investigated period. Moreover, the VIKOR method results comply with acceptable advantage and acceptable stability in decision-making requirements. The dispersion of performance rankings for Nordic and Baltic countries over the period from 2013 to 2017 are presented in Figure 10.6.

As presented in figure 10.6, the performance rankings of the countries remained stable over the investigated period. Nordic countries generally performed better in comparison to Baltic countries. Latvia and Lithuania were ranked last over the investigated period.

CONCLUSION AND POLICY IMPLICATIONS

An increase in global warming triggered by CO_2 emissions has adverse effects on economies all over the world (Lin et al., 2020). Thus, economic performance and climate change mitigation are interrelated. Maintaining economic performance and climate change mitigation at the same time is crucial for developed countries, since the scales of economies are affected by climate change. In this context, we aim to evaluate economic performance and climate change mitigation with an integrated model. We employ the VIKOR method to evaluate the economic performance and climate change mitigation of Nordic and Baltic countries over the period from 2013 to 2017.

According to the results of the VIKOR method, Sweden has been found to be the best-performing country over the investigated period. Two Baltic countries, namely Latvia and Lithuania, have been found as the last-place alternatives in the rankings. Nordic countries generally performed better in comparison to Baltic countries. This result is due to the investments carried out by Nordic countries. Nordic countries are pioneers in renewable energy technology and use (Sovacool, 2017). Renewable energy technologies have

Table 10.4 VIKOR Method Results and Rankings for Nordic and Baltic Countries

Countries	2013 Q_i Values	2013 Ranking	2014 Q_i Values	2014 Ranking	2015 Q_i Values	2015 Ranking	2016 Q_i Values	2016 Ranking	2017 Q_i Values	2017 Ranking
Sweden	0.0000	1	0.0000	1	0.0000	1	0.0000	1	0.0000	1
Iceland	0.2754	2	0.3018	3	0.2895	3	0.2770	2	0.2746	2
Estonia	0.3217	3	0.3070	4	0.2715	2	0.3249	3	0.2860	3
Finland	0.4024	4	0.2927	2	0.2925	4	0.3998	4	0.3456	4
Norway	0.4750	5	0.4762	5	0.4907	5	0.4892	5	0.4711	5
Denmark	0.6641	6	0.5419	6	0.5639	6	0.6470	6	0.5855	6
Latvia	0.7371	7	0.7689	7	0.7336	7	0.7464	7	0.7307	7
Lithuania	1.0000	8	1.0000	8	1.0000	8	1.000	8	1.0000	8

Source: Authors' calculations.

a crucial role in maintaining a sustainable energy structure in developed countries (Miremadi et al., 2019). Sweden has gone through enormous energy transformation in recent years especially for solar and wind power (Glaa & Mignon, 2020). Sweden has presented the necessary circumstances for local foundations. Long-term strict devotion to this goal is the key element of renewable energy success. The hydropower option is an important renewable energy alternative in mountainous areas, and the coastal regions are effectively utilized for wind energy in Sweden (Kooij et al., 2018). One of the most important strategies when dealing with GHG emissions is fossil fuel tax, which dates back to the early twentieth century in Sweden (Shmelev & Speck, 2018). The Swedish economy has grown 10% over the last decade. It is highly important to increase climate change mitigation performance at the same time as economic growth. Energy efficiency and a low carbon economy have been increasingly important due to increases in global warming (Johansson & Ranius, 2019). Apart from environmental concerns, energy efficiency is also a key factor in maintaining economic growth. Being energy efficient means less expenditure on energy-oriented goods, and it enables countries to invest in other macroeconomic indicators. Energy efficiency may prevent shortages of energy sources as a result of high energy prices (Ohene-Asare et al., 2020). Nordic countries generally have a high energy efficiency level, which provides them with economic growth (Irandoust, 2016).

Another country that has been successful in economic performance and climate change mitigation is Iceland, which possesses a high amount of RESs (Shafiei et al., 2018). Iceland is one of the top RES users, thanks to its hydropower and geothermal sources. Iceland produces nearly all its electricity from RESs. This independence in electricity has enabled Iceland to provide competitive energy prices, and this advantage has led to more power plants (Cook et al., 2016).

The Baltic country with the top performance is Estonia, which primarily invests in wind farms. Estonia is a quite advantageous country, as it has large areas for wind farms on sea coasts. Since Estonia gained its independence, it has separated from the European electric grid. Estlink, which is a cable line connected to secure the energy supply of Nordic and Baltic countries, contributes to the energy security of this region. New cables are planned to further integrate Baltic countries with Nordic countries. This investment is an important milestone for leadership in terms of climate change mitigation in Nordic and Baltic countries among European Union (EU) countries (Crandall, 2014). Estonia stimulates renewable energy through subsidies and premium tariffs (Proskurina et al., 2016). One of the Nordic countries most successful at mitigating climate change is Finland. Finland is a pioneer country in bioenergy use thanks to its large forest areas. Finland aims to increase the share of bioenergy in transportation, too (Zakeri et al., 2015). Strict regulations to reduce emissions are key elements for maintaining climate change mitigation efforts (Shakeel et al., 2017).

The Baltic states, consisting of Latvia, Lithuania, and Estonia, were formerly ruled by the Soviet regime. They gained their independence in the 1990s, and this separation has brought independence to the economy and environmental issues. For instance, they secured an energy structure eliminating the demand for gas from Russia (Štreimikienė et al., 2019). Lithuania is in the last place in the ranking among Nordic and Baltic countries, but, even though the country performs relatively poorly in comparison to other countries in the analysis, Lithuania has increased its share of production and consumption of RES. Economic growth has remained steady over the last years in Lithuania. Novel investments are necessary, however, to reach the 2030 and 2050 European clean environment targets. In this regard, wind power is considered to have a key role in mitigating climate change in Lithuania (Gaigalis & Katinas, 2020). Today, Baltic countries have achieved their 2020 RES target (Štreimikienė et al., 2019). They are more successful than most of the EU countries in terms of economic performance and climate change mitigation. They demonstrate relatively low performance, however, vis-à-vis Nordic countries. The devotion of Nordic countries to RES depends on policy instruments, such as carbon taxes, feed-in tariffs, investment grants, and electricity certificates. Another key point in climate change mitigation is research and development for new energy technologies (Irandoust, 2016). In this regard, the countries with low climate change mitigation performance should adopt these policies to increase their performance level. This performance increase will also reflect in the form of economic growth.

As a result, we may conclude that MCDM methods are precious decision science techniques for analyzing climate change mitigation and the economics of the energy sector. Our empirical results and policy implications may provide remarkable results for both decision-makers and stakeholders in terms of climate change mitigation, particularly in the field of the energy sector.

NOTES

1. Figures 10.1 through 10.6 available online at https://rowman.com/ISBN/9781793625021/Sustainable-Engineering-for-Life-Tomorrow.

REFERENCES

Ardakani, S. R., Hossein, S. M., & Aslani, A. (2018). Statistical approaches to forecasting domestic energy consumption and assessing determinants: The case of Nordic countries. *Strategic Planning for Energy and the Environment, 38*(1), 26–71.

Arya, V., & Kumar, S. (2020). A new picture fuzzy information measure based on Shannon entropy with applications in opinion polls using extended VIKOR–TODIM approach. *Computational and Applied Mathematics, 39*(3), 1–24.

Aslani, A., Naaranoja, M., & Wong, K. F. V. (2013). Strategic analysis of diffusion of renewable energy in the Nordic countries. *Renewable and Sustainable Energy Reviews, 22*, 497–505.

Babatunde, K. A., Begum, R. A., & Said, F. F. (2017). Application of computable general equilibrium (CGE) to climate change mitigation policy: A systematic review. *Renewable and Sustainable Energy Reviews, 78*, 61–71.

Baležentis, T., Li, T., Streimikiene, D., & Baležentis, A. (2016). Is the Lithuanian economy approaching the goals of sustainable energy and climate change mitigation? Evidence from DEA-based environmental performance index. *Journal of Cleaner Production, 116*, 23–31.

Cardenas, L. M., Franco, C. J., & Dyner, I. (2016). Assessing emissions–mitigation energy policy under integrated supply and demand analysis: The Colombian case. *Journal of Cleaner Production, 112*, 3759–3773.

Chen, A. A., Stephens, A. J., Koon, R. K., Ashtine, M., & Koon, K. M. K. (2020). Pathways to climate change mitigation and stable energy by 100% renewable for a small island: Jamaica as an example. *Renewable and Sustainable Energy Reviews, 121*, 109671.

Christis, M., Athanassiadis, A., & Vercalsteren, A. (2019). Implementation at a city level of circular economy strategies and climate change mitigation—The case of Brussels. *Journal of Cleaner Production, 218*, 511–520.

Cook, D., Davíðsdóttir, B., & Kristófersson, D. M. (2016). Energy projects in Iceland—Advancing the case for the use of economic valuation techniques to evaluate environmental impacts. *Energy Policy, 94*, 104–113.

Crandall, M. (2014). Soft security threats and small states: The case of Estonia. *Defence Studies, 14*(1), 30–55.

Cucchiella, F., D'Adamo, I., Gastaldi, M., & Miliacca, M. (2018). Efficiency and allocation of emission allowances and energy consumption over more sustainable European economies. *Journal of Cleaner Production, 182*, 805–817.

Elum, Z. A., & Momodu, A. S. (2017). Climate change mitigation and renewable energy for sustainable development in Nigeria: A discourse approach. *Renewable and Sustainable Energy Reviews, 76*, 72–80.

Eurostat (EU Statistical Bureau). Retrieved June 20, 2020, from https://ec.europa.eu/eurostat/web/energy/data/database

Eurostat Website. Retrieved June 20, 2020, from https://ec.europa.eu/energy/topics/energy-strategy/previous-energy-strategies_en#the-2020-energy-strategy-2010-

Fedajev, A., Stanujkic, D., Karabašević, D., Brauers, W. K., & Zavadskas, E. K. (2020). Assessment of progress towards "Europe 2020" strategy targets by using the MULTIMOORA method and the Shannon Entropy Index. *Journal of Cleaner Production, 244*, 118895.

Gaigalis, V., & Katinas, V. (2020). Analysis of the renewable energy implementation and prediction prospects in compliance with the EU policy: A case of Lithuania. *Renewable Energy, 151*, 1016–1027.

Glaa, B., & Mignon, I. (2020). Identifying gaps and overlaps of intermediary support during the adoption of renewable energy technology in Sweden—A conceptual framework. *Journal of Cleaner Production, 261*, 121178.

Gökgöz, F. (2010). Measuring the financial efficiencies and performances of Turkish funds. *Acta Oeconomica, 60*(3), 295–320.

Halkos, G., & Petrou, K. N. (2019). Assessing 28 EU member states' environmental efficiency in national waste generation with DEA. *Journal of Cleaner Production, 208*, 509–521.

Irandoust, M. (2016). The renewable energy-growth nexus with carbon emissions and technological innovation: Evidence from the Nordic countries. *Ecological Indicators, 69*, 118–125.

Jahan, A., Mustapha, F., Ismail, M. Y., Sapuan, S. M., & Bahraminasab, M. (2011). A comprehensive VIKOR method for material selection. *Materials & Design, 32*(3), 1215–1221.

Johansson, J., & Ranius, T. (2019). Biomass outtake and bioenergy development in Sweden: The role of policy and economic presumptions. *Scandinavian Journal of Forest Research, 34*(8), 771–778.

Johnsson, F., Kjärstad, J., & Rootzén, J. (2019). The threat to climate change mitigation posed by the abundance of fossil fuels. *Climate Policy, 19*(2), 258–274.

Khanali, M., Kokei, D., Aghbashlo, M., Nasab, F. K., Hosseinzadeh-Bandbafha, H., & Tabatabaei, M. (2020). Energy flow modeling and life cycle assessment of apple juice production: Recommendations for renewable energies implementation and climate change mitigation. *Journal of Cleaner Production, 246*, 118997.

Kim, Y. K. (2013). Household debt, financialization, and macroeconomic performance in the United States, 1951–2009. *Journal of Post Keynesian Economics, 35*(4), 675–694.

Kooij, H. J., Oteman, M., Veenman, S., Sperling, K., Magnusson, D., Palm, J., & Hvelplund, F. (2018). Between grassroots and treetops: Community power and institutional dependence in the renewable energy sector in Denmark, Sweden and the Netherlands. *Energy Research & Social Science, 37*, 52–64.

Leimbach, M., Kriegler, E., Roming, N., & Schwanitz, J. (2017). Future growth patterns of world regions—A GDP scenario approach. *Global Environmental Change, 42*, 215–225.

Lin, X., Zhu, X., Han, Y., Geng, Z., & Liu, L. (2020). Economy and carbon dioxide emissions effects of energy structures in the world: Evidence based on SBM-DEA model. *Science of the Total Environment, 729*, 138947.

Løken, E. (2007). Use of multicriteria decision analysis methods for energy planning problems. *Renewable and Sustainable Energy Reviews, 11*(7), 1584–1595.

Maghsoodi, A. I., Abouhamzeh, G., Khalilzadeh, M., & Zavadskas, E. K. (2018). Ranking and selecting the best performance appraisal method using the MULTIMOORA approach integrated Shannon's entropy. *Frontiers of Business Research in China, 12*(1), 2.

Manish, S., Pillai, I. R., & Banerjee, R. (2006). Sustainability analysis of renewables for climate change mitigation. *Energy for Sustainable Development, 10*(4), 25–36.

Mavi, R. K., Goh, M., & Mavi, N. K. (2016). Supplier selection with Shannon entropy and fuzzy TOPSIS in the context of supply chain risk management. *Procedia: Social and Behavioral Sciences, 235*, 216–225.

Mi, Z. F., Wei, Y. M., He, C. Q., Li, H. N., Yuan, X. C., & Liao, H. (2017). Regional efforts to mitigate climate change in China: A multi-criteria assessment approach. *Mitigation and Adaptation Strategies for Global Change, 22*(1), 45–66.

Miremadi, I., Saboohi, Y., & Arasti, M. (2019). The influence of public R&D and knowledge spillovers on the development of renewable energy sources: The case of the Nordic countries. *Technological Forecasting and Social Change, 146,* 450–463.

Miskinis, V., Galinis, A., Konstantinaviciute, I., Lekavicius, V., & Neniskis, E. (2020). Comparative analysis of energy efficiency trends and driving factors in the Baltic States. *Energy Strategy Reviews, 30,* 100514.

Ohene-Asare, K., Tetteh, E. N., & Asuah, E. L. (2020). Total factor energy efficiency and economic development in Africa. *Energy Efficiency, 13*(6), 1177–1194.

Opricovic, S. (1998). Multicriteria optimization of civil engineering systems. *Faculty of Civil Engineering, Belgrade, 2*(1), 5–21.

Ploskas, N., & Papathanasiou, J. (2019). A decision support system for multiple criteria alternative ranking using TOPSIS and VIKOR in fuzzy and nonfuzzy environments. *Fuzzy Sets and Systems, 377,* 1–30.

Pohekar, S. D., & Ramachandran, M. (2004). Application of multi-criteria decision making to sustainable energy planning—A review. *Renewable and Sustainable Energy Reviews, 8*(4), 365–381.

Proskurina, S., Sikkema, R., Heinimö, J., & Vakkilainen, E. (2016). Five years left—How are the EU member states contributing to the 20% target for EU's renewable energy consumption; the role of woody biomass. *Biomass and Bioenergy, 95,* 64–77.

San Cristóbal, J. R. (2011). Multi-criteria decision-making in the selection of a renewable energy project in Spain: The Vikor method. *Renewable Energy, 36*(2), 498–502.

Santamouris, M. (2016). Innovating to zero the building sector in Europe: Minimising the energy consumption, eradication of the energy poverty and mitigating the local climate change. *Solar Energy, 128,* 61–94.

Sarkodie, S. A., Adams, S., & Leirvik, T. (2020). Foreign direct investment and renewable energy in climate change mitigation: Does governance matter? *Journal of Cleaner Production, 263,* 121262.

Shafiei, E., Davidsdottir, B., Fazeli, R., Leaver, J., Stefansson, H., & Asgeirsson, E. I. (2018). Macroeconomic effects of fiscal incentives to promote electric vehicles in Iceland: Implications for government and consumer costs. *Energy Policy, 114,* 431–443.

Shakeel, S. R., Takala, J., & Zhu, L. D. (2017). Commercialization of renewable energy technologies: A ladder building approach. *Renewable and Sustainable Energy Reviews, 78,* 855–867.

Shannon, C. E. (1948). A mathematical theory of communication. *Bell System Technical Journal, 27*(3), 379–423.

Sharifi, A. (2020). Co-benefits and synergies between urban climate change mitigation and adaptation measures: A literature review. *Science of the Total Environment, 750*(50) 141642.

Shmelev, S. E., & Speck, S. U. (2018). Green fiscal reform in Sweden: Econometric assessment of the carbon and energy taxation scheme. *Renewable and Sustainable Energy Reviews, 90,* 969–981.

Siksnelyte, I., Zavadskas, E. K., Bausys, R., & Streimikiene, D. (2019). Implementation of EU energy policy priorities in the Baltic Sea Region countries: Sustainability assessment based on neutrosophic MULTIMOORA method. *Energy Policy, 125,* 90–102.

Škrinjarić, T. (2020). Empirical assessment of the circular economy of selected European countries. *Journal of Cleaner Production, 255,* 120246.

Sovacool, B. K. (2017). Contestation, contingency, and justice in the Nordic low-carbon energy transition. *Energy Policy, 102,* 569–582.

Štreimikienė, D., Mikalauskienė, A., Atkočiūnienė, Z., & Mikalauskas, I. (2019). Renewable energy strategies of the Baltic states. *Energy & Environment, 30*(2), 363–381.

Tan, R., & Lin, B. (2018). What factors lead to the decline of energy intensity in China's energy intensive industries? *Energy Economics, 71,* 213–221.

Velasco-Fernández, R., Giampietro, M., & Bukkens, S. G. (2018). Analyzing the energy performance of manufacturing across levels using the end-use matrix. *Energy, 161,* 559–572.

Wang, B., Wang, Q., Wei, Y. M., & Li, Z. P. (2018). Role of renewable energy in China's energy security and climate change mitigation: An index decomposition analysis. *Renewable and Sustainable Energy Reviews, 90,* 187–194.

Wang, D., Du, Z., & Wu, H. (2020). Ranking global cities based on economic performance and climate change mitigation. *Sustainable Cities and Society, 62,* 102395.

Wang, T. C., & Lee, H. D. (2009). Developing a fuzzy TOPSIS approach based on subjective weights and objective weights. *Expert Systems with Applications, 36*(5), 8980–8985.

Wennersten, R., Sun, Q., & Li, H. (2015). The future potential for carbon capture and storage in climate change mitigation—An overview from perspectives of technology, economy and risk. *Journal of Cleaner Production, 103,* 724–736.

Wood, N., & Roelich, K. (2019). Tensions, capabilities, and justice in climate change mitigation of fossil fuels. *Energy Research & Social Science, 52,* 114–122.

Xu, F., Liu, J., Lin, S., & Yuan, J. (2017). A VIKOR-based approach for assessing the service performance of electric vehicle sharing programs: A case study in Beijing. *Journal of Cleaner Production, 148,* 254–267.

Yang, W., Xu, K., Lian, J., Ma, C., & Bin, L. (2018). Integrated flood vulnerability assessment approach based on TOPSIS and Shannon entropy methods. *Ecological Indicators, 89,* 269–280.

Yazdani, M., Zarate, P., Zavadskas, E. K., & Turskis, Z. (2019). A Combined Compromise Solution (CoCoSo) method for multi-criteria decision-making problems. *Management Decision, 57*(3): 1–19.

Zakeri, B., Syri, S., & Rinne, S. (2015). Higher renewable energy integration into the existing energy system of Finland—Is there any maximum limit? *Energy, 92,* 244–259.

Zheng, X., Streimikiene, D., Balezentis, T., Mardani, A., Cavallaro, F., & Liao, H. (2019). A review of greenhouse gas emission profiles, dynamics, and climate change mitigation efforts across the key climate change players. *Journal of Cleaner Production, 234,* 1113–1133.

Index

About the Editors

Jacqueline A. Stagner is the undergraduate programs coordinator of engineering at the University of Windsor. She is also an adjunct graduate faculty member in the Department of Mechanical, Automotive and Materials Engineering. She co-advises students in the sustainability and renewable energy areas, in the Turbulence & Energy Laboratory. She has co-edited two volumes.

David S. K. Ting is the founder of the Turbulence & Energy Laboratory, University of Windsor. As a professor in the Department of Mechanical, Automotive and Materials Engineering, he supervises students primarily on energy and flow turbulence. To date, Professor Ting has co/supervised over 75 graduate students, coauthored more than 140 journal papers, authored 4 textbooks and coedited 10 volumes.

About the Contributors

Yomna K. Abdallah is an assistant professor at Universitat Internacional de Catalunya, School of Architecture. She obtained PhD in biodigital architecture in 2020. Her PhD was in a joint/channel supervision collaboration between four universities (Universitat Internacional de Catalunya. Biodigital Architecture, University of Granada. Biochemistry, Helwan University. Interior Design, Cairo University. Microbiology), her research on embedding bioactive agents in the built environment, for producing bioelectricity was funded by the Ministry of Higher Education of the Egyptian Government, and the Erasmus Mundus Mobility and Training Program Erasmus+. Abdallah is now conducting her second PhD in bioengineering in the Faculty of Health Science in Universitat Internacional de Catalunya to develop new biomaterials.

Julia Ann Baker works as a quality management system engineer at Danfoss Power Solutions since January 2019. Before that she had the chance to work as the head of projects and Development of Nordic Platform, at Siemens Gamesa as an engineering trainee and for a longer period at Kvik A/S as a project coordinator. She studied at Aarhus University, earning a bachelor of engineering (BEng) in the field of global management and manufacturing between 2011 and 2015 and MSc in engineering in technology-based business development graduating in 2018.

Figen Balo received BSc, MSc, and PhD degrees in mechanical engineering from Firat University. She worked at the Ministries of Environment-Urbanization and Education. She is currently professor, head of department in industrial engineering at Firat University. She has published sixty-two papers in scientific journals, sixteen book chapters, and ninety-two research papers in conferences. Her main research areas include renewable energy, building insulation materials containing natural sources, optimization of energy efficiency at buildings, and MCDM for the best selection of equipment at renewable energy systems.

Rupp Carriveau is the director of the Environmental Energy Institute and the Turbulence and Energy Lab at the University of Windsor. His research activities focus on energy systems futures. Dr. Carriveau serves on the Editorial Boards of Wind Engineering, *Advances in Energy Research*, and the *International Journal of Sustainable Energy*. He recently guest-edited special editions of *Energies* and *The Journal of Energy Storage*. Professor Carriveau was a recent recipient of the University Scholar Award and has acted as a Research Ambassador for the Council of Ontario Universities. Dr. Carriveau is a founder of the Offshore Energy and Storage Society (OSES) and recently co-chaired OSES2018 Ningbo China, and OSES2019 Brest France. Professor Carriveau is chair of the IEEE Ocean Energy Technology Committee and was just named to Canada's Clean50 2020 for his contributions to clean capitalism.

Roberto Francisco Coelho received the BS, MS, and PhD degrees from the Federal University of Santa Catarina (UFSC), Florianopolis, SC, Brazil, in 2006, 2008, and 2013, respectively. He is currently an adjunct professor with the Department of Electrical and Electronics Engineering, UFSC. To date, he has co/supervised over twenty-eight graduate students, co-authored more than thirty journal papers, and authored one book chapter. His main areas of research include power converters, control-oriented modeling, control techniques applied to power converters, maximum power point tracker systems, and grid-connected photovoltaic inverters.

Peter Enevoldsen is an associate professor at the Centre for Energy Technologies, at the Department of Business Development and Technology, Aarhus University and mainly focuses on the development of new and innovative energy systems for businesses and consumers, primarily in the areas of wind energy, energy efficiency, energy policy, and hydrogen. His detailed work on wind farms in forested areas has upgraded the potentials of well-known software tools in the field of wind resource assessment, such as WAsP and WindPRO, tools used to improve the energy efficiency of wind farms.

Alberto T. Estévez is an architect, with double PhD & studies (Architecture and Art History). He heads his own office of architecture and design in Barcelona. He was teaching architectural design, theory, and arts in different universities, until he found the ESARQ (UIC) in 1996, where he is professor of architecture, after being its first director. He founded as director the UIC PhD programs in architecture, and in 2000 the Biodigital Architecture Master's Degree and the Genetic Architectures Research Group & Office, included in the iBAG-UIC Barcelona (Institute for Biodigital Architecture & Genetics) of which he is the director. He was also vice chancellor/general manager of UIC Barcelona (Universitat Internacional de Catalunya). He has written more than 200 publications, participating in numerous exhibitions and conferences around the world.

Fazıl Gökgöz received his BSc and MSc degrees in engineering and earned MBA and PhD in management with Superior Achievement Award. He served as acting head of Privatization Project Group and associate in numerous M&A operations at the Privatization Administration of Turkey. He served as member of board directors and member of auditing board at various state-owned companies. He is a full-time professor in Ankara University Faculty of Political Sciences. He served as rectorate management coordinator and vice dean of faculty of political sciences at Ankara University. He teaches quantitative methods at undergraduate and graduate levels and has carried out numerous international academic publications on energy, finance, and quantitative methods.

Neveen M. Khalil is full-time professor of microbiology in Cairo University, Faculty of Science, Department of Botany and Microbiology. Khalil specializes in biotechnological applications of fungal strains, focusing mainly on environmental, agricultural, and industrial applications of filamentous fungal strains. Khalil has numerous published papers in indexed/Scopus journals, focusing on the study of different strains *Aspergillus* and their possible applications in industry.

Kheir Al-Kodmany is a professor of sustainable spatial planning and urban design at the University of Illinois at Chicago (UIC). He has published 6 major books and more than 100 papers. He was invited to lecture in dozens of world's renowned universities and is member of the editorial board of twenty journals. Before joining the UIC faculty, he worked for the Chicago firm Skidmore, Owings & Merrill.

Rui Kou is a postdoctoral researcher at the University of California, San Diego. He earned a PhD in structural engineering from the University of California, San Diego in 2020, after earning a master's degree in material science from the University of California, San Diego in 2016 and a bachelor's degree in welding science and technology from Harbin Institute of Technology in 2015. His research interests include intelligent materials/

structure, energy conversion materials, nanomaterials, and advanced infra-structural materials.

Denizar Cruz Martins (Senior Member, IEEE) was born in São Paulo, SP, Brazil in 1955. He received the BS and MS degrees in electrical engineering from the Federal University of Santa Catarina (UFSC), Florianopolis, SC, Brazil, in 1978 and 1981, respectively, and PhD in electrical engineering from the Polytechnic National Institute of Toulouse, Toulouse, France, in 1986. He is currently a titular professor with the Department of Electrical and Electronics Engineering, UFSC. His research interests include DC–DC and DC–AC converters, high-frequency soft commutation, power factor correction, and grid-connected photovoltaic systems.

Lindsay Miller is an adjunct professor in civil engineering and instructor in technical communications at the University of Windsor. Lindsay has notable research experience in energy systems with a focus on renewable energy, energy finance, and the environmental and economic impacts of energy technologies and sources. As a steering committee member for Women and Inclusivity in Sustainable Energy Research (WISER) and Electricity Human Resources Canada (EHRC)'s Women in Leadership, Lindsay is passionate about promoting diversity and inclusion in the energy sector and in the field of engineering. Lindsay has several publications in her research areas and, along with her colleagues, has hosted many workshops, conferences, and events across North America.

Negin Minaei is a visiting scholar at the City Institute at the York University in Toronto. As a university faculty and lecturer, she has taught and researched in different universities and countries for two decades including the University of Windsor (Canada), RAU (UK), Shandong Agricultural University (China), IAU and Bahonar University (Iran), and Bauhaus Dessau (Germany). The main areas of her research have been sustainable smart cities, impacts of IT, TC and advanced technologies on global cities, and future cities, sustainable urbanism and transportation, cognitive maps, navigation, GPS and transport modes, Echo-Tech architecture using active/passive solar design systems, and zero-energy buildings.

Sarah Nichol is a graduate student at the University of Windsor currently completing the qualifications to receive an MASc civil and environmental engineering degree. She is also a member of the Environmental Energy Institute and Turbulence and Energy Lab at the University of Windsor. Her research focuses on innovations in energy systems, with a focus on offshore energy systems.

Lenon Schmitz was born in Blumenau, SC, Brazil, in 1990. He received his BS, MS, and PhD degrees in electrical engineering from the Federal University of Santa Catarina, Florianopolis, SC, Brazil, in 2013, 2015, and 2020, respectively. He served as a substitute professor in the Department of Electrical and Electronics Engineering at UFSC, from 2018 to 2019, and is currently a researcher at the Power Electronics Institute at UFSC. His interest research areas include high-efficiency power converters, design optimization techniques, and grid-connected photovoltaic systems.

Zhaoru Shang is a PhD candidate at the University of California, San Diego. He earned master's degree in material science and engineering from the University of California, San Diego in 2019 and bachelor's degree in materials science and engineering from the University of Chinese Academy of Sciences. His research interests include energy conversion materials, physical modeling, and molecular dynamics simulations.

Mostafa M. Sobhy is a teaching assistant at Arab Academy for Science, Technology & Maritime Transport, Egypt. Sobhy has been working with renowned architectural firms in Egypt such as Algorithm and MUD studio. He has also participated in many workshops of parametric design and digital fabrication with Sapienza University in Italy.

Lutfu S. Sua obtained his PhD in production and operations management from the University of Mississippi, USA in 2005. He also has an MBA from Troy University, USA. He has been teaching as a professor of operations management for the past fifteen years. He also minored in economics and management information systems. His primary research area is the quantitative analysis and mathematical programming. Other research interests are renewable energy systems, Industry 4.0, and supply chain management. He has more than 120 publications in international academic journals and conference proceedings within these fields.

Diaa El Deen M. Tantawy is assistant professor in Helwan University, Faculty of Applied Arts, Department of Interior Design and Furniture, and MSA, Faculty of Art and Design. Tantawy is specialized in digital and parametric architecture. He has numerous publications in indexed journals focusing on the use of digital design and fabrication tools in the architectural realm.

Marco Samuel Moesby Tinggaard is the CTO & cofounder of PESITHO, a dynamic start-up that opened in February 2019. He holds a degree from Aarhus University as a business development engineer (2013–2016) and an MSc in engineering in technology-based business development from the same university, graduating in 2018.

George Xydis had the opportunity to work in the wind sector together with developers, utilities, constructors, universities, and research institutes. He used to work as a Wind Projects Development Coordinator at Iberdrola Renewables, as a Wind Project Developer at Vector Hellenic Windfarms S.A., and as a Researcher at the Centre for Electric Power and Energy, Technical University of Denmark. He is now an associate professor at the Centre for Energy Technologies, Department of Business Development and Technology, Aarhus University. George is an adjunct lecturer at Johns Hopkins University at Energy Policy and Climate program. He teaches 425.624—Wind Energy: Science, Technology and Policy. He has coauthored more than 100 publications in international journals, conferences, books and book chapters, and serves as an associate editor in Frontiers in Energy Systems & Policy and in Energy & Environment. He holds a PhD from National Technical University of Athens, and a degree in mechanical engineering from Aristotle University of Thessaloniki, Greece.

Engin Yalçın was born in 1990. He completed BA in management from Gazi University in 2014 and MBA from Pamukkale University in 2017. He is currently pursuing his PhD at Ankara University Department of Management and is a research assistant at Ankara University Social Sciences Institute.

Haozhe Yi is a graduate researcher at the University of California, San Diego. He is currently a PhD candidate in structural engineering from the University of California, San Diego. He has a master's degree in civil engineering from the University of Virginia in 2018 and a bachelor's degree in civil engineering from Dongguan University of Technology in 2015. His research interests include intelligent materials/structure, structure design, and advanced infrastructural materials.

Unal Yilmaz was born in Malatya, Turkey. He is the founder of the Asri Engineering Co., Turkey. The main task of his company is projecting and establishment of renewable energy systems for companies or individuals. His main research areas are renewable energy, optimization of energy efficiency at buildings, and simulation of renewable energy systems.

Mingyuan Zhang is a graduate researcher at the University of Washington. He is currently pursuing his PhD in materials science and engineering at the University of Washington after earning his master's degree in materials science and engineering from the University of California, San Diego in 2020 and his bachelor degree in materials science and engineering from Beijing University of Technology in 2017. His research interests include active biochemical separation systems and energy storage materials.

Ying Zhong is currently an assistant professor at Department of Mechanical Engineering at University of South Florida (USF). Before joining USF in 2019, she was a postdoctoral fellow in Department of Structural Engineering at University of California at San Diego (UCSD). She received her PhD in materials science and engineering from UCSD in 2017. In UCSD, her research focused on the development of energy efficiency improvement technologies for buildings based on advanced designs and materials. As the PI of USF GREEN Lab (Green Research for Energy-Efficient iNnovations) (www.usfgreen.com) at USF, she is conducting research in advanced disinfection technology, intelligent sensing systems, and energy-efficient systems.

www.ingramcontent.com/pod-product-compliance
Lightning Source LLC
Chambersburg PA
CBHW050640280326
41932CB00015B/2724